THEORY OF ALGEBRAIC INTEGERS

Theory of Algebraic Integers

Richard Dedekind

Translated and introduced by John Stillwell

CAMBRIDGE
UNIVERSITY PRESS

Published by the Press Syndicate of the University of Cambridge
The Pitt Building, Trumpington Street, Cambridge CB2 1RP
40 West 20th Street, New York, NY 10011-4211, USA
10 Stamford Road, Oakleigh, Melbourne 3166, Australia

First published in French 1877

First published in English 1996

Library of Congress cataloging in publication data
Dedekind, Richard, 1831-1916.
Theory of algebraic integers / Richard Dedekind;
translated and with an introduction by John Stillwell
p.   cm.
Includes bibliographical references and index.
ISBN 0-521-56518-9 (pbk.)
1. Algebraic number theory. 2. Integral representations.
I. Title
QA247.D43   1996
512′.74–dc20   96-1601 CIP

British Library cataloguing in publication data available

ISBN 0 521 56518 9   paperback

Transferred to digital printing 2004

Contents

Contents

Part one

Translator's introduction

Translator's introduction

0.1 General remarks

Dedekind's invention of ideals in the 1870s was a major turning point in the development of algebra. His aim was to apply ideals to number theory, but to do this he had to build the whole framework of commutative algebra: fields, rings, modules and vector spaces. These concepts, together with groups, were to form the core of the future abstract algebra. At the same time, he created algebraic number theory, which became the temporary home of algebra while its core concepts were growing up. Algebra finally became independent in the 1920s, when fields, rings and modules were generalised beyond the realm of numbers by Emmy Noether and Emil Artin. But even then, Emmy Noether used to say "Es steht schon bei Dedekind" ("It's already in Dedekind"), and urged her students to read all of Dedekind's works in ideal theory.

Today this is still worthwhile, but not so easy. Dedekind wrote for an audience that knew number theory – especially quadratic forms – but not the concepts of ring, field or module. Today's readers probably have the opposite qualifications, and of course most are not fluent in German and French. In an attempt to overcome these problems, I have translated the most accessible of Dedekind's works on ideal theory, *Sur la Théorie des Nombres Entiers Algébriques*, Dedekind (1877), which he wrote to explain his ideas to a general mathematical audience. This memoir shows the need for ideals in a very concrete case, the numbers $m + n\sqrt{-5}$ where $m, n \in \mathbb{Z}$, before going on to develop a general theory and to prove the theorem on unique factorisation into prime ideals.

The algebraic integers in Dedekind's title are a generalisation of the ordinary integers – created in response to certain limitations of classical number theory. The ordinary integers have been studied since ancient times, and their basic theory was laid down in Euclid's *Elements* (see Heath (1925)) around 300 BC. Yet even ancient number theory contains problems not solvable by Euclid's methods. Sometimes it is necessary to use irrational numbers, such as $\sqrt{2}$, to answer questions about the ordinary integers. A famous example is the so-called Pell equation

$$x^2 - cy^2 = 1$$

where c is a nonsquare integer and the solutions x, y are required to be integers. Solutions for certain values of c were known to the ancients, but the complete solution was not obtained until Lagrange (1768) related the equation to the continued fraction expansion of $\sqrt{c}$. He also showed that each solution is obtained from a certain "minimum" solution (x_0, y_0) by the formula

$$x_k + y_k\sqrt{c} = \pm(x_0 + y_0\sqrt{c})^k.$$

The irrational numbers $x_k + y_k\sqrt{c}$ in this formula are examples of *algebraic integers*, which are defined in general to be roots of equations of the form

$$\alpha^n + a_{n-1}\alpha^{n-1} + \cdots + a_1\alpha + a_0 = 0$$

where $a_{n-1}, \ldots, a_0$ are ordinary integers, that is, $a_{n-1}, \ldots, a_0 \in \mathbb{Z}$.

Algebraic integers are so called because they share some properties with the ordinary integers. In particular, they are closed under sum, difference and product, and the rational algebraic integers are just the ordinary integers (for more details, see 0.6.2 and §13 of Dedekind's memoir). Because of the second fact, the ordinary integers are also known as *rational integers*. The first fact implies that the algebraic integers form a ring. However, we are not interested in the ring of all algebraic integers so much as rings like

$$\mathbb{Z}[\sqrt{2}] = \{x + y\sqrt{2} : x, y \in \mathbb{Z}\}$$

and

$$\mathbb{Z}[i] = \{x + yi : x, y \in \mathbb{Z}\}.$$

In general we use the notation $\mathbb{Z}[\alpha]$ to denote the closure of the set $\mathbb{Z} \cup \{\alpha\}$ under $+$, $-$ and $\times$. The reason for working in rings $\mathbb{Z}[\alpha]$ is that they more closely resemble $\mathbb{Z}$, and hence are more likely to yield information about $\mathbb{Z}$.

Any ring R of algebraic integers includes $\mathbb{Z}$, so theorems about $\mathbb{Z}$ may be obtainable as special cases of theorems about R (we shall see several examples later). However, useful theorems about R are provable only when R has all the basic properties of $\mathbb{Z}$, in particular, *unique prime factorisation*. This is not always the case. $\mathbb{Z}[\sqrt{-5}]$ is the simplest example where unique prime factorisation fails, and this is why Dedekind studies it in detail. His aim is to recapture unique prime factorisation by extending the concept of integer still further, to certain sets of algebraic integers he calls *ideals*. This works only if the size of R is limited in some way. The ring A of all algebraic integers is "too big" because it includes $\sqrt{\alpha}$ along with each algebraic integer α. This gives the factorisation $\alpha = \sqrt{\alpha}\sqrt{\alpha}$ and hence "primes" do not exist in A, let alone unique prime factorisation.

Dedekind found the appropriate "small" rings R in algebraic number fields of finite degree, each of which has the form $\mathbb{Q}(\alpha)$, where α is an algebraic integer. $\mathbb{Q}(\alpha)$ denotes the closure of $\mathbb{Q} \cup \{\alpha\}$ under $+,-,\times$ and $\div$ (by a nonzero number), and each $\mathbb{Q}(\alpha)$ has its own integers, which factorise into primes. In particular, $\mathbb{Z}[\sqrt{-5}]$ is the ring of integers of $\mathbb{Q}(\sqrt{-5})$, and $6 = 2 \times 3$ is a prime factorisation of 6. Not *the* prime factorisation, alas, because $6 = (1+\sqrt{-5})(1-\sqrt{-5})$ is also a factorisation into primes (see 0.4.5). However, unique prime factorisation is regained when one passes to the ideals of $\mathbb{Z}[\sqrt{-5}]$, and Dedekind generalises this to any $\mathbb{Q}(\alpha)$. The result is at last a theory of algebraic integers capable of yielding information about ordinary integers.

A lot of machinery is needed to build this theory, but Dedekind explains it well. Suffice to say that fields, rings and modules arise very naturally as sets of numbers closed under the basic operations of arithmetic. Fields are closed under $+$, $-$, $\times$ and $\div$, rings are closed under $+$, $-$ and $\times$, while modules are closed under $+$ and $-$. The term "ring" was actually introduced by Hilbert (1897); Dedekind calls them "domains" here, and I have thought it appropriate to retain this terminology, since these particular rings are prototypes of what are now called *Dedekind domains*. Dedekind presumably chose the name "module" because a module M is something for which "congruence modulo M" is meaningful. His name for field, Körper (which also means "body" in German), was chosen to describe "a system with a certain completeness, fullness and self-containedness; a naturally unified, organic whole", as he explained in his final exposition of ideal theory, Dedekind (1894), §160.

What Dedekind does not explain is where $\mathbb{Z}[\sqrt{-5}]$ comes from, and why it is important in number theory. This is understandable, because his first version of ideal theory was a supplement to Dirichlet's number theory lectures, *Vorlesungen über Zahlentheorie* (Dirichlet (1871)). In the present memoir he also refers to the *Vorlesungen* frequently, so his original audience was assumed to have a good background in number theory, and particularly the theory of quadratic forms. Such a background is less common today, but is easy and fun to acquire. Even experts may be surprised to learn how far back the story goes. The specific role of $\sqrt{-5}$ can be traced back to the anomalous behaviour of the quadratic form $x^2 + 5y^2$, first noticed by Fermat, and later explained in different ways by Lagrange, Gauss and Kummer. But the reason for Fermat's interest in $x^2 + 5y^2$ goes back much further, perhaps to the prehistory of mathematics in ancient Babylon. Let us begin there.

0.2 Squares

0.2.1 Pythagorean triples

Integers a, b, c such that

$$a^2 + b^2 = c^2$$

are one of the oldest treasures of mathematics. Such numbers occur as the sides of right-angled triangles, and they may even have been used to construct right angles in ancient times. They are called Pythagorean triples after Pythagoras, but they were actually discovered independently in several different cultures. The Babylonians were fascinated by them as early as 1800 BC, when they recorded fifteen of them on a tablet now known as Plimpton 322 (see Neugebauer and Sachs (1945)). Pythagorean triples other than the simplest ones (3,4,5), (5,12,13) or (8,15,17) are not easily found by trial and error, so the Babylonians probably knew a general formula such as

$$a = 2uv, \quad b = u^2 - v^2, \quad c = u^2 + v^2$$

which yields an unlimited supply of Pythagorean triples by substituting different integers for u, v.

The general solution of $a^2 + b^2 = c^2$ is in fact

$$a = 2uvw, \quad b = (u^2 - v^2)w, \quad c = (u^2 + v^2)w,$$

as may be found in Euclid's *Elements* Book X (lemma after Proposition 29). A key statement in Euclid's proof is: *if the product of relatively prime integers is a square, then the integers themselves are squares.*

Euclid first used the general formula for Pythagorean triples in his theory of irrational numbers, and it is in a different book from his theory of integers. The assumption that relatively prime integers are squares when their product is a square is justified by a long chain of proposi-tions, stretching over several books of the *Elements*. However, a direct justification is possible from his theory of integer divisibility, which is in Book VII. This theory is fundamental to the theory of ordinary integers, and also the inspiration for Dedekind's theory of ideals, so we should re-call its main features before going any further. Among other things, it identifies the important but elusive role of primes.

0.2.2 Divisors and prime factorisation

An integer m *divides* an integer n if $n = ml$ for some integer l. We also say that m is a *divisor* of n, or that n is a *multiple* of m. An integer p whose only divisors are ± 1 and $\pm p$ is called a *prime*, and any integer can be factorised into a finite number of primes by successively finding divisors unequal to ± 1 but of minimal absolute value. However, it is not obvious that each factorisation of an integer n involves the *same* set of primes. There is conceivably a factorisation of some integer

$$n = p_1 p_2 \cdots p_i = q_1 q_2 \cdots q_j$$

into primes $p_1, p_2, \ldots, p_i$ and $q_1, q_2, \ldots, q_j$ respectively, where one of the primes p is different from all the primes q.

Nonunique prime factorisation is ruled out by the following proposi-tion of Euclid (*Elements*, Book VII, Proposition 30).

Prime divisor property. *If p is prime and p divides the product ab of integers a, b, then p divides a or p divides b.*

An interesting aspect of the proof is its reliance on the concept of *greatest common divisor* (gcd), particularly the fact that

$$\gcd(a, b) = ua + vb \quad \text{for some integers } u, \ v.$$

The set $\{ua + vb : u, v \in \mathbb{Z}\}$ is in fact an *ideal*, and unique prime factori-sation is equivalent to the fact that this ideal consists of the multiples of one of its members, namely $\gcd(a, b)$.

It should be mentioned that Euclid proves only the prime divisor prop-erty, not unique prime factorisation. In fact its first explicit statement and proof are in Gauss (1801), the *Disquisitiones Arithmeticae*, article 16. As we shall see, this is possibly because Gauss was first to recognise generalisations of the integers for which unique prime factorisation is not valid.

0.2.3 Irrational numbers

As everybody knows, Pythagorean triples also have significance as the sides of right-angled triangles. In any right-angled triangle, the side lengths a, b, c satisfy

$$a^2 + b^2 = c^2$$

whether or not a, b and c are integers (Pythagoras' theorem). Hence it is tempting to try to interpret a right-angled triangle as a Pythagorean triple by choosing the unit of length so that a, b and c all become integer lengths. Pythagoras or one of his followers made the historic discovery that this is not always possible. The simplest counterexample is the triangle with sides 1, 1, $\sqrt{2}$. It is impossible to interpret this triangle as a Pythagorean triple because $\sqrt{2}$ is not a rational number.

A proof of this fact, which also proves the irrationality of $\sqrt{3}$, $\sqrt{5}$, $\sqrt{6}$ and so on, uses unique prime factorisation to see that each prime appears to an even power in a square. Then the equation

$$2n^2 = m^2$$

is impossible because the prime 2 occurs an odd number of times in the prime factorisation of $2n^2$, and an even number of times in the prime factorisation of m^2.

The irrationality of $\sqrt{2}$ led the Greeks to study the so-called *Pell equation*

$$x^2 - 2y^2 = 1.$$

They found it could used to approach $\sqrt{2}$ rationally, via increasingly large integer solutions, x_n, y_n. Since $x_n^2 - 2y_n^2 = 1$, the quotient x_n/y_n necessarily tends to $\sqrt{2}$. The general Pell equation

$$x^2 - cy^2 = 1, \quad \text{where } c \text{ is a nonsquare integer,}$$

can similarly be used to approach the irrational number $\sqrt{c}$. This equation later proved fruitful in many other ways; Dedekind even used it to prove the irrationality of $\sqrt{c}$ (Dedekind (1872), Section IV).

0.2.4 Diophantus

The equations $a^2 + b^2 = c^2$ and $x^2 - 2y^2 = 1$ are examples of what we now call *Diophantine equations*, after Diophantus of Alexandria. Diophantus lived sometime between 150 AD and 350 AD and wrote a collection of books on number theory known as the *Arithmetica* (Heath (1910)). They consist entirely of equations and ingenious particular solutions. The term "Diophantine" refers to the type of solution sought: either rational or integer. For Diophantus it is usually a rational solution, but for some equations, such as the Pell equation, the integer solutions are of more interest. The Pell equation was actually not studied by Diophantus, but he mentioned an integer solution to another remarkable equation: the solution $x = 5$, $y = 3$ of $y^3 = x^2 + 2$. (See 0.4.1 for the astonishing sequel to this solution.)

Although all Diophantus' solutions are special cases, they usually seem chosen to illustrate general methods. Euler (1756) went so far as to say

Nevertheless, the actual methods that he uses for solving any of his problems are as general as those in use today ... there is hardly any method yet invented in this kind of analysis not already traceable to Diophantus. (Euler *Opera Omnia* I,2, p. 429–430.)

And if anyone would know, Euler would. The first mathematician to understand Diophantus properly was Fermat (1601–1665), but his comments were as cryptic as the *Arithmetica* itself. Euler spent about 40 years, off and on, reading between the lines of Fermat and Diophantus, until he could reconstruct most their methods and prove their theorems. We shall study the connection between Diophantus, Fermat and Euler more thoroughly later, but one example is worth mentioning here. It shows how much theory can be latent in a single numerical fact.

In the *Arithmetica*, Book III, Problem 19, Diophantus remarks

65 is naturally divided into two squares in two ways, namely into $7^2 + 4^2$ and $8^2 + 1^2$, which is due to the fact that 65 is the product of 13 and 5, each of which is the sum of two squares.

It appears from this that Diophantus is aware of the identity

$$(a^2 + b^2)(c^2 + d^2) = (ac \pm bd)^2 + (ad \mp bc)^2$$

though he makes no such general statement. However, Fermat saw much deeper than this. Noticing, with Diophantus, that the identity reduces the representations of a number as a sum of two squares to the representations of its prime factors, his comment on the problem is:

A prime number of the form $4n+1$ is the hypotenuse of a right-angled triangle [that is, a sum of squares] in one way only . . . If a prime number which is the sum of two squares be multiplied by another prime number which is also the sum of two squares, the product will be the sum of two squares in two ways. (Heath (1910), p. 268.)

The restriction to primes of the form $4n + 1$ is understandable because a prime $p \neq 2$ cannot be the sum of two squares unless it is of the form $4n+1$ (by a congruence mod 4 argument). But Fermat's claim that any prime $p = 4n + 1$ is a sum of two squares comes right out of the blue. No one knows how he proved it and the first known proof is due to Euler (1756). As we shall see later, Lagrange, Gauss and Dedekind all used this theorem of Fermat to test the strength of new methods in number theory.

0.3 Quadratic forms

0.3.1 Fermat

Unlike Euclid or Diophantus, Fermat never wrote a book. His reputation rests on a short manuscript containing his discovery of coordinate geometry (independent of Descartes), his letters, and his marginal notes on Diophantus. He took up number theory only in his late 30s, and left only one reasonably complete proof, in the posthumously published Fermat (1670). However, it is a beautiful piece of work, and fully establishes his credentials as both an innovator and a student of the ancients. It also has a place in our story, as an application of Pythagorean triples, and as the first proven instance of Fermat's last theorem. Fermat's proof shows that there are no positive integers x, y, z such that $x^4 + y^4 = z^4$, by showing that there are not even positive integers x, y, z such that $x^4 + y^4 = z^2$. It turns out to be the only instance of Fermat's last theorem with a really elementary proof, involving just Euclid's theory of divisibility.

The argument is by contradiction, and the gist of it is as follows (omitting mainly routine checks that certain integers are relatively prime).

Suppose that there are positive integers x, y, z such that $x^4+y^4 = z^2$, or in other words, $(x^2)^2 + (y^2)^2 = z^2$. This says that x^2, y^2, z is a Pythagorean triple, which we can take to be primitive, hence there are integers u, v such that

$$x^2 = 2uv, \quad y^2 = u^2 - v^2, \quad z = u^2 + v^2,$$

by Euclid's formulas (0.2.1). The middle equation says that v, y, u is also a Pythagorean triple, and it is also primitive, hence there are integers s, t such that

$$v = 2st, \quad y = s^2 - t^2, \quad u = s^2 + t^2.$$

This gives

$$x^2 = 2uv = 4st(s^2 + t^2),$$

so the relatively prime integers s, t and $s^2 + t^2$ have product equal to the square $(x/2)^2$. It follows that each is itself a square, say

$$s = x_1^2, \quad t = y_1^2, \quad s^2 + t^2 = z_1^2,$$

and hence

$$x_1^4 + y_1^4 = z_1^2.$$

Thus we have found another sum of two fourth powers equal to a square, and by retracing the argument we find that the new square z_1^2 is *smaller* than the old, z^2, but still nonzero. By repeating the process we can therefore obtain an infinite descending sequence of positive integers, which is a contradiction.   □

Fermat called the method used in this proof *infinite descent*, and used it for many of his other theorems. He claimed, for example, to have proved that any prime of the form $4n + 1$ is a sum of two squares by supposing $p = 4n + 1$ to be a prime not the sum of two squares, and finding a smaller prime with the same property. However, it is very hard to see how to make the descent in this case. Euler (1749) found a proof only after several years of effort. In 0.3.4 we shall see an easier proof of the two squares theorem due to Lagrange. Lagrange's proof does use another famous theorem of Fermat, but it is the easy one known as Fermat's "little" theorem: *for any prime number p, and any integer $a \not\equiv 0 \pmod{p}$, we have $a^{p-1} \equiv 1 \pmod{p}$* (Fermat (1640b)).

The proof of Fermat's little theorem most likely used by Fermat uses induction on a and the fact that a prime p divides each of the binomial coefficients

$$\binom{p}{i} = \frac{p(p-1)(p-2)\cdots(p-i+1)}{i!} \quad \text{for} \quad 1 \leq i \leq p - 1,$$

as is clear from the fact that p is a factor of the numerator but not of the denominator. This proof implicitly contains the "mod p binomial theorem",

$$(a + b)^p \equiv a^p + b^p \pmod{p},$$

which has its uses elsewhere (see for example Gauss's proof of quadratic reciprocity in 0.5.4).

The proof more often seen today is based on that of Euler (1761), which implicitly uses the group properties of multiplication mod p, particularly the idea of multiplicative inverses. An integer a is nonzero mod p if $\gcd(a, p) = 1$, in which case $1 = ar + ps$ for some integers r, s by the Euclidean algorithm. The number r is called a *multiplicative inverse* of a $(\bmod\ p)$ since $ar \equiv 1$ $(\bmod\ p)$. It follows that mod p multiplication by a nonzero a is invertible, and in particular the set $\{a \times 1, a \times 2, \ldots, a \times (p-1)\}$ is the same set $(\bmod\ p)$ as the set $\{1, 2, \ldots, p-1\}$. Hence each set has the same product mod p,

$$a \times 2a \times \cdots \times (p-1)a \equiv 1 \times 2 \times \cdots \times (p-1) \quad (\bmod\ p),$$

and cancellation of $1, 2, \ldots, p-1$ from both sides (which is permissible, since $1, 2, \ldots, p-1$ have inverses) gives Fermat's little theorem:

$$a^{p-1} \equiv 1 \ (\bmod\ p).$$

0.3.2 The grit in the oyster

The mathematical pearl that is Dedekind's theory of ideals grew in response to a tiny irritant, the anomalous behaviour of the quadratic form $x^2 + 5y^2$. Between 1640 and 1654 Fermat discovered three beautiful theorems about the representation of odd primes p by the forms $x^2 + y^2$, $x^2 + 2y^2$ and $x^2 + 3y^2$ for integer values of x and y (the first prompted by Diophantus' remark on sums of squares, as mentioned in 0.2.4):

Theorem 1. $p = x^2 + y^2 \ \Leftrightarrow\ p \equiv 1 \ (\bmod\ 4)$ (Fermat (1640c))

Theorem 2. $p = x^2 + 2y^2 \ \Leftrightarrow\ p \equiv 1$ or $3 \ (\bmod\ 8)$ (Fermat (1654))

Theorem 3. $p = x^2 + 3y^2 \ \Leftrightarrow\ p \equiv 1 \ (\bmod\ 3)$ (Fermat (1654))

Fermat thought he could prove these theorems, and he was probably right, as proofs eventually published by Euler were based on Fermat's method of infinite descent. Since $x^2 + 4y^2 = x^2 + (2y)^2$, which is "of the form" $x^2 + y^2$, the next theorem should be about $x^2 + 5y^2$. This is the grit in the oyster. Fermat was unable to prove a theorem about primes of the form $x^2 + 5y^2$, and could only conjecture the following less satisfying fact.

If two primes which end in 3 or 7 and surpass by 3 a multiple of 4 are multiplied, then their product will be composed of a square and the quintuple of another square. (Fermat (1654).)

Since numbers that end in 3 or 7 are of the form $10n + 3$ or $10n + 7$, and such a number is also of the form $4m + 3$ if and only if it is of the form $20k + 3$ or $20k + 7$, Fermat's conjecture can be restated as follows:

Fermat's conjecture. *If two primes are of the form $20k + 3$ or $20k + 7$ then their product is of the form $x^2 + 5y^2$.*

This is very puzzling. What about primes of the form $x^2 + 5y^2$? There *are* some, such as

$$29 = 3^2 + 5 \times 2^2,$$
$$41 = 6^2 + 5 \times 1^2,$$
$$61 = 4^2 + 5 \times 3^2,$$

and furthermore, they lie in the classes $20k + 1$ and $20k + 9$ that seem conspicuously absent from the conjecture above. This situation begs for an explanation. As Weil (1974) says:

When there is something that is really puzzling and cannot be understood, it usually deserves the closest attention because some time or other some big theory will emerge from it.

Euler (1744) found another clue to the puzzle. He noticed two "faces" to the behaviour of $x^2 + 5y^2$ – sometimes it represents a prime p, sometimes $2p$ – and he conjectured that the following holds for all prime values p.

Euler's conjecture. $p = x^2 + 5y^2 \Leftrightarrow p \equiv 1$ or $9 \pmod{20}$
$\qquad\qquad\qquad 2p = x^2 + 5y^2 \Leftrightarrow p \equiv 3$ or $7 \pmod{20}$.

0.3.3 Reduction of forms

The first to account rigorously for the two-faced behaviour of $x^2 + 5y^2$ was Lagrange (1773). Studying the general question of which integers n could be represented by a given quadratic form $ax^2 + bxy + cy^2$ (where $a, b, c \in \mathbb{Z}$), he had the very fruitful idea of finding those forms $a'x'^2 + b'x'y' + c'y'^2$ equivalent to $ax^2 + bxy + cy^2$ via a change of variables.

The substitution

$$x' = \alpha x + \beta y,$$
$$y' = \gamma x + \delta y$$

maps $\mathbb{Z} \times \mathbb{Z}$ into $\mathbb{Z} \times \mathbb{Z}$ provided $\alpha, \beta, \gamma, \delta \in \mathbb{Z}$ and it is one-to-one provided there is an inverse substitution

$$x = \alpha'x' + \beta'y',$$
$$y = \gamma'x' + \delta'y'.$$

In this case

$$\begin{bmatrix} \alpha & \beta \\ \gamma & \delta \end{bmatrix} \begin{bmatrix} \alpha' & \beta' \\ \gamma' & \delta' \end{bmatrix} = \begin{bmatrix} 1 & 0 \\ 0 & 1 \end{bmatrix},$$

since the product of a substitution and its inverse is the identity. There-fore, taking determinants of both sides,

$$\begin{vmatrix} \alpha & \beta \\ \gamma & \delta \end{vmatrix} \begin{vmatrix} \alpha' & \beta' \\ \gamma' & \delta' \end{vmatrix} = 1.$$

Finally, since the determinants on the left are integers and the only integer divisors of 1 are ± 1, it follows that the invertible substitutions for which the pairs (x', y') run through $\mathbb{Z} \times \mathbb{Z}$ are precisely those with $\alpha, \beta, \gamma, \delta \in \mathbb{Z}$ and determinant $\alpha\delta - \beta\gamma = \pm 1$. Such substitutions are now called *unimodular*.

The result $a'x'^2 + b'x'y' + c'y'^2$ of a unimodular substitution $\alpha'x' + \beta'y'$ for x and $\gamma'x' + \delta'y'$ for y in $ax^2 + bxy + cy^2$ is therefore a form that takes the same values as $ax^2 + bxy + cy^2$. Forms transformable into each other by unimodular substitutions are equivalent, as we would say, because the unimodular substitutions form a group. Lagrange observed that equivalent forms have the same *discriminant*

$$D = b^2 - 4ac = b'^2 - 4a'c',$$

as can be checked by computing $b'^2 - 4a'c'$ and using $\alpha\delta - \beta\gamma = \pm 1$. (Incidentally, the old term for the discriminant of a quadratic form was *determinant*. I have retained this term in the translation of Dedekind's memoir because he refers to a slightly different definition, due to Gauss.) Observing the invariance of the discriminant is a first step towards decid-ing equivalence of forms. To go further we need to answer the question: how many inequivalent forms have the same discriminant?

Lagrange found a way to answer this question for forms with negative discriminant. He showed that any $ax^2 + bxy + cy^2$ can be transformed into an equivalent form $a'x'^2 + b'x'y' + c'y'^2$ that is *reduced* in the sense that $|b'| \le a' \le c'$.

It follows that

$$-D = 4a'c' - b'^2 \ge 4a'^2 - a'^2 = 3a'^2$$

and therefore, in the case of negative discriminant, only finitely many values of the integers a', b' and hence c', can occur in reduced forms. For any particular $D < 0$ it is then possible to work out the inequivalent reduced forms of discriminant D. The number of them is called the *class number $h(D)$*. The first few calculations yield the following results.

All forms with discriminant -4 *are equivalent to* $x^2 + y^2$, *hence*
$$h(-4) = 1.$$
All forms with discriminant -8 *are equivalent to* $x^2 + 2y^2$, *hence*
$$h(-8) = 1.$$
All forms with discriminant -12 *are equivalent to* $x^2 + 3y^2$, *hence*
$$h(-12) = 1.$$

But *there are two inequivalent reduced forms with discriminant* -20, *the forms* $x^2 + 5y^2$ *and* $2x^2 + 2xy + 3y^2$, *so* $h(-20) = 2$. Aha! There *is* something different about $x^2 + 5y^2$! Perhaps the previously invisible companion $2x^2 + 2xy + 3y^2$ accounts for its "two-faced" behaviour. Before following up this suggestion, however, let us see how Lagrange used the uniqueness of the form $x^2 + y^2$ to prove the two squares theorem.

0.3.4 Lagrange's proof of the two squares theorem

Given a prime $p = 4n + 1$, it suffices to find any form with discriminant -4 that represents p, since Lagrange has shown we can transform this form into $x^2 + y^2$. It suffices in turn to find an integer m such that p divides $m^2 + 1$, because in this case the form $px^2 + 2mxy + \frac{m^2+1}{p}y^2$ has integer coefficients, its discriminant is -4, and of course it takes the value p for $x = 1$, $y = 0$. (The use of this particular form is a simplification of Lagrange's argument due to Gauss (1801), article 182).

This is where it is crucial that p be of the form $4n + 1$ and prime. By Fermat's little theorem, p then divides $z^{4n} - 1 = (z^{2n} - 1)(z^{2n} + 1)$ for any integer z relatively prime to p. Thus p will divide $z^{2n} + 1$, which is $m^2 + 1$ with $m = z^n$, provided z is chosen so that p does *not* divide $z^{2n} - 1$. In terms of congruences, we want to choose one of the $4n$ nonzero values of z (mod p) so that $z^{2n} - 1$ is nonzero (mod p). This is possible by an earlier theorem of Lagrange (1770): a congruence of degree q modulo a prime p has at most q solutions.

A modern proof of this theorem (not much different from Lagrange's own proof) uses the fact that the nonzero integers mod p have multiplicative inverses. It follows that the congruence classes mod p form a field, and one can show as in classical algebra that each root z_i of a polynomial $p(z)$ corresponds to a factor $(z - z_i)$ of $p(z)$. Hence there cannot be more roots than the degree of $p(z)$.

It follows in the present case that we have at least $4n - 2n = 2n$ values of z which make $z^{2n} - 1$ nonzero mod p, and hence p divides $z^{2n} + 1 = m^2 + 1$ as required.

0.3.5 Primitive roots and quadratic residues

Lagrange's result that an nth degree congruence mod p has at most n different solutions has another important consequence in mod p arithmetic – the existence of primitive roots. We digress a little further here to discuss this concept, since it will be important later, and it also throws more light on the two squares theorem.

An integer a is called a *primitive root* mod p if

$$a^{p-1} \equiv 1 \pmod{p},$$
$$a^q \not\equiv 1 \pmod{p} \quad \text{for } 1 \le q < p-1.$$

More generally, if we define the *order* of element a mod p to be the least positive n such that $a^n \equiv 1 \pmod{p}$, then a primitive root is an element of order $p - 1$. Existence of a primitive root means that the group of nonzero congruence classes is cyclic. Euler conjectured that a primitive root exists for each p, but was unable to prove it. The first proof was given by Gauss (1801), article 55. Like all proofs since, the primitive root is not constructed explicitly – rather, its existence is shown by counting the number of possible solutions of a congruence.

The quickest proof looks at elements of relatively prime order, and first shows that the least common multiple l of the orders of nonzero integers mod p is itself the order of some element, a. Since l is the least common multiple we then have

$$x^l \equiv 1 \pmod{p} \quad \text{for} \quad x = 1, 2, \ldots, p-1.$$

If $l < p - 1$ this is a congruence of degree l with more than l solutions, contrary to Lagrange's result. Hence in fact $l = p - 1$, which means a is a primitive root mod p. □

The existence of a primitive root a mod p means that

$$\{1, a, a^2, \ldots, a^{p-2}\} = \{1, 2, \ldots, p-1\}$$

This makes many facts about multiplication mod p quite transparent. For example, *exactly half the integers* $1, 2, \ldots, p-1$ *are squares, mod* p. Indeed, the even powers $1, a^2, a^4, \ldots$ of a primitive root are squares mod p, and conversely.

The traditional term for squares mod p is *quadratic residues* mod p. Quadratic residues arise naturally in the study of quadratic forms, and their fundamental theorem, called "quadratic reciprocity", was conjectured by Euler. Like the existence of primitive roots (also conjectured by Euler), it was first proved by Gauss. We shall say more about this in Section 0.5. Two important preliminaries to the general discussion of squares mod p are the following:

Euler's criterion. *m is a square mod* $p \Leftrightarrow m^{\frac{p-1}{2}} \equiv 1 \pmod{p}$

If a is a primitive root mod p,

$$m \text{ is square mod } p \Rightarrow m = a^{2j} \quad \text{for some } j$$
$$\Rightarrow m^{\frac{p-1}{2}} = a^{j(p-1)} \equiv 1 \pmod{p},$$
$$m \text{ is nonsquare mod } p \Rightarrow m = a^{2j+1} \quad \text{for some } j$$
$$\Rightarrow m^{\frac{p-1}{2}} = a^{j(p-1)+\frac{p-1}{2}} \equiv a^{\frac{p-1}{2}} \equiv -1 \pmod{p}.$$

Quadratic character of -1. *The number* -1 *is a square mod* $p \Leftrightarrow$ $p = 4n + 1$ *for some* n.

Notice that this statement strengthens the result used in the proof of the two squares theorem, that $p = 4n + 1$ divides some $m^2 + 1$. It also follows easily from the existence of primitive roots:

$$-1 \text{ is a square mod } p \Leftrightarrow x^2 \equiv -1 \pmod{p} \quad \text{for some } x$$
$$\Leftrightarrow x \text{ has order } 4 \pmod{p}$$
$$\Leftrightarrow x = a^{\frac{p-1}{4}} \quad \text{where } a \text{ is a primitive root}$$
$$\Leftrightarrow p = 4n + 1 \quad \text{for some } n.$$

0.3.6 Composition of forms

The second reduced form of discriminant -20, namely $2x^2 + 2xy + 3y^2$, accounts for much of the anomalous behaviour of $x^2 + 5y^2$. The primes not represented by $x^2 + 5y^2$, namely those in the classes $20n + 3$ and $20n + 7$, are represented by $2x^2 + 2xy + 3y^2$, for example

$$3 = 2 \times 0^2 + 2 \times 0 \times 1 + 3 \times 1^2,$$
$$7 = 2 \times 1^2 + 2 \times 1 \times 1 + 3 \times 1^2,$$
$$23 = 2 \times (-2)^2 + 2 \times (-2) \times 3 + 3 \times 3^2.$$

Lagrange (1773) proved this, and also Euler's conjecture (0.3.2) that all primes p in the classes $20n + 1$ and $20n + 9$ are of the form $x^2 + 5y^2$. He used these two theorems to establish the conjectures of Fermat and Euler (0.3.2) about products and doubles of primes in the classes $20n + 3$ and $20n + 7$, with the help of another observation: the product of two numbers of the form $2x^2 + 2xy + 3y^2$ is a number of the form $x^2 + 5y^2$. This crucial observation is based on the following algebraic identity:

$$(2x^2 + 2xy + 3y^2)(2x'^2 + 2x'y' + 3y'^2) = X^2 + 5Y^2$$

where

$$X = 2xx' + xy' + yx' - 2yy',$$
$$Y = xy' + yx' + yy'.$$

Such an identity can be checked mechanically by multiplying out both sides, but how did Lagrange find it in the first place? Probably by past experience with products of forms, some of which were known much earlier. We have already seen one known to Diophantus (0.2.4):

$$(x^2 + y^2)(x'^2 + y'^2) = (xx' - yy')^2 + (xy' + yx')^2$$

It has a generalisation

$$(x^2 + cy^2)(x'^2 + cy'^2) = (xx' - cyy')^2 + c(xy' + yx')^2$$

discovered by the Indian mathematician Brahmagupta around 600 AD. (See Colebrooke (1817), p. 363, and Weil (1984), p. 14.)

This would have been easy for Lagrange to derive using the complex factors

$$(x^2 + cy^2)(x'^2 + cy'^2) = (x + y\sqrt{-c})(x - y\sqrt{-c})(x' + y'\sqrt{-c})(x' - y'\sqrt{-c}),$$

and pairing the first factor with the third, and the second with the fourth. Lagrange's own identity can be derived quite mechanically from

$$(x^2 + 5y^2)(x'^2 + 5y'^2) = (xx' - 5yy')^2 + 5(xy' + yx')^2$$

(Brahmagupta's identity for $c = 5$), and

$$2x^2 + 2xy + 3y^2 = 2\left[\left(x + \frac{y}{2}\right)^2 + 5\left(\frac{y}{2}\right)^2\right]$$

(the result of completing the square on $2x^2 + 2xy + 3y^2$). Use the latter to rewrite each of $2x^2 + 2xy + 3y^2$ and $2x'^2 + 2x'y' + 3y'^2$, in the form $2[X^2 + 5Y^2]$, multiply them out using Brahmagupta's identity, then absorb the factors of 2.

There is a related identity for the product of the two different forms of discriminant -20:

$$(x^2 + 5y^2)(2x'^2 + 2x'y' + 3y'^2) = 2X^2 + 2XY + 3Y^2$$

where

$$X = xx' - yx' - 3yy',$$
$$Y = xy' + 2yx' + yy'$$

and it can be derived in a similar way. These identities show that the forms $x^2 + 5y^2$ and $2x^2 + 2xy + 3y^2$ are "closed under products" in a certain sense. The product operation is known as *composition of forms*.

Legendre (1798) managed to show, in fact, that any two quadratic forms with the same discriminant could be "composed" in this fashion. Something very interesting was going on, but what?

0.3.7 The class group

Composition of forms came on the scene decades before the axiomatic properties of abstract structures, such as groups, were considered in mathematics. Legendre had found a set (the forms with fixed negative discriminant) and a "product" operation on it (composition) but there was no reason to expect the operation to have simple or interesting structural properties. All he could was draw up "multiplication tables" for the forms with particular discriminants. For example, if we take the forms with discriminant -20,

$$A = x^2 + 5y^2,$$
$$B = 2x^2 + 2xy + 3y^2,$$

the table would be

	A	B
A	A	B
B	B	A

because $AA = A$ by Brahmagupta's identity, $BB = A$ by Lagrange's identity, and $AB = BA = B$ by the last identity of 0.3.6.

A more complicated example actually given by Legendre was for the forms of discriminant -164, rewritten by Cox (1989)) p. 42 as:

$$A = x^2 + 41y^2,$$
$$B = 2x^2 + 2xy + 21y^2,$$
$$C = 5x^2 + 6xy + 10y^2,$$
$$D = 3x^2 + 2xy + 14y^2,$$
$$E = 6x^2 + 2xy + 7y^2.$$

Legendre's multiplication table for these forms can then be written:

	A	B	C	D	E
A	A	B	C	D	E
B	B	A	C	E	D
C	C	C	A or B	D or E	D or E
D	D	E	D or E	A or C	B or C
E	E	D	D or E	B or C	A or C

To us, of course, it is rather disturbing to see $AC = BC$ when $A \neq B$, and worse still that some of the products have two values. The two-valued entries are due to an ambiguity of sign in Legendre's definition of composition. Ambiguity would be avoided in any modern attempt to define an operation on a set, and it *was* avoided in the next study of quadratic forms, by Gauss (1801) (the *Disquisitiones Arithmeticae*).

Gauss's analysis of forms under composition is amazingly modern in some ways. He defined composition unambiguously, showed that it is well-defined on equivalence classes of forms and showed, in effect, that the equivalence classes of forms constitute an abelian group under composition. It is now called the *class group*. Gauss even came close to finding a decomposition of the class group into cyclic factors.

Yet all this was accomplished with definitions and proofs so cumbersome it took 70 years for the rest of the world to understand them. With Gauss's definition of composition, for example, it is a major problem to prove that composition is associative. Even the statement of associativity in the *Disquisitiones* is clumsy, as if Gauss had not really grasped what associativity is about:

If the form F is composed of the forms f, f'; the form $\mathfrak{F}$ from F and f''; the form F' from f, f''; the form $\mathfrak{F}'$ from F' and f'; then the forms $\mathfrak{F}$, $\mathfrak{F}'$ will be ... equivalent. (Gauss (1801), article 240.)

It is only by using the commutativity of composition (which is mercifully obvious) that this statement can be rewritten in a form recognisable as associativity, namely

$$f''(ff') = (f''f)f'.$$

The proof is monstrous. It requires the derivation of 37 equations, most of which Gauss leaves to the reader. This put the subject of composition of forms out of the reach of most mathematicians until Dirichlet and Dedekind simplified it enough to make associativity obvious. A polished exposition was eventually given by Dedekind in his Supplement X to the 2nd edition of Dirichlet's *Vorlesungen über Zahlentheorie* (Dirichlet (1871)).

However, the approach via algebraic identities, no matter how slick, was becoming irrelevant by this time. The abstract structure of the class group was seen as more important, and more usable, since facts could be deduced from it without reference to the definition of composition. For example, Kronecker (1870) showed that the decomposition of the class group into cyclic factors follows purely from axioms for finite abelian groups (which he was the first to state). Dedekind's Supplement X to Dirichlet (1871) was in effect the swansong of the old theory of composition of forms, because in the same work he showed how to rebuild the class group on a simpler and more general basis, the theory of algebraic integers. The first important examples of such integers were studied by Lagrange and Euler.

0.4 Quadratic integers

0.4.1 The need for generalised "integers"

We have already seen how convenient it is to use factorisations involving square roots to prove identities about integers, such as

$$(x^2 + cy^2)(x'^2 + cy'^2) = (xx' - cyy')^2 + c(xy' + yx')^2.$$

Lagrange (1768) and Euler (1770) began using complex or irrational factorisations to find integer solutions of equations. The most interesting and provocative example is the Euler (1770) proof of the following claim of Fermat (1657): 27 *is the only cube that exceeds a square by* 2. This is equivalent to saying that $y = 3$, $x = 5$ is the only positive integer solution of the equation

$$y^3 = x^2 + 2.$$

Euler's solution is breathtaking, even if not exactly rigorous. He factorises the right hand side into $(x - \sqrt{-2})(x + \sqrt{-2})$, and proceeds to treat $x - \sqrt{-2}$ and $x - \sqrt{-2}$ *as if they were integers*. He assumes they have gcd 1, and therefore, since their product is a cube, they are cubes themselves. This presumably means that

$$x + \sqrt{-2} = (a + b\sqrt{-2})^3 \quad \text{for some} \quad a, b \in \mathbb{Z}$$
$$= a^3 + 3a^2b\sqrt{-2} + 3ab^2(-2) + b^3(-2)\sqrt{-2}.$$

and therefore

$$x = a^3 - 6ab^2,$$
$$1 = 3a^2b - 2b^3$$

equating real and imaginary parts. But

$$1 = 3a^2b - 2b^3 = b(3a^2 - b^2)$$

only if $b = \pm 1$ and $a = \pm 1$, since 1 and -1 are the only integer divisors of 1. This gives $x = 5$ as the only positive integer solution for x. Q.E.D!

Euler gave several examples of this kind, generally splitting quadratics into irrational complex factors and treating the factors as integers. Why is this permissible, if indeed it is? To answer this question we need to recall how ordinary integers behave, particularly as divisors.

The behaviour of the ordinary integers is ruled by unique prime factorisation, which in turn depends on the prime divisor property: *a prime divides a product ab only if it divides one of a, b* (0.2.2). For example, this is crucial in proving that relatively prime integers are squares when their product is a square (0.2.1). To justify similar propositions about "quadratic integers" such as $x + \sqrt{-2}$, we have to decide which of them are "primes" and then see whether they have a prime divisor property like the ordinary integers. This is easiest when the quadratic integers in question have a "Euclidean algorithm", because one can then follow the trail blazed by Euclid in his proof of the prime divisor property.

The first to carry out such a program was Gauss (1832), who studied the divisibility properties of the numbers $x + y\sqrt{-1}$, where $x, y \in \mathbb{Z}$. These numbers are now known as the *Gaussian integers*. They are the simplest kind of quadratic integers, and they tie up nicely with quadratic forms and some other threads in our story, so it is worth looking at them first.

0.4.2 Gaussian integers

Making the usual abbreviation i for $\sqrt{-1}$, we denote the set of Gaussian integers $x + yi$ by $\mathbb{Z}[i]$. The first step towards unique prime factorisation in $\mathbb{Z}[i]$ is to show that primes exist. We can do this immediately with the help of the *norm*, a measure of size in $\mathbb{Z}[i]$ introduced by Gauss. The norm of $x + yi$, $N(x + yi)$, is defined by

$$N(x + yi) = |x + yi|^2 = x^2 + y^2,$$

and has the multiplicative property

$$N((x + yi)(x' + y'i)) = N(x + yi)N(x' + y'i),$$

because

$$(xx' - yy')^2 + (xy' + yx')^2 = (x^2 + y^2)(x'^2 + y'^2),$$

by Diophantus' identity (0.2.4). It follows that any divisor $x + yi$ of $X + Yi$ has $N(x + yi) \leq N(X + Yi)$ and therefore, since the norm is a positive integer, each Gaussian integer of norm > 1 has a divisor $x + yi$ of minimal norm > 1.

We call any such divisor a *Gaussian prime* because it is not the product of Gaussian integers of smaller norm. By repeatedly removing prime divisors, we obtain a factorisation of any Gaussian integer into Gaussian primes. Obviously these factors are determined only up to factors of norm 1, the so-called *units* 1, -1, i, $-i$. This is similar to the situation in $\mathbb{Z}$, where prime factors are determined only up to the unit factors ± 1. The real problem in $\mathbb{Z}[i]$, as in $\mathbb{Z}$, is to establish the prime divisor property: *if a prime divides a product then it divides one of the factors.*

Euclid's proof of the prime divisor property for $\mathbb{Z}$ (0.2.2) involves his algorithm for finding the gcd by repeated subtraction, which terminates because it yields a decreasing sequence of positive integers. The proof for $\mathbb{Z}[i]$ is similar, but the algorithm for gcd requires repeated division with remainder, which terminates because it yields a sequence of remainders that decrease in norm. The decrease is guaranteed by the following *division property of* $\mathbb{Z}[i]$: *for any Gaussian integers α and $\beta \neq 0$ there are Gaussian integers μ and ρ ("quotient" and "remainder") such that*

$$\alpha = \mu\beta + \rho \quad where \quad 0 \leq |\rho| < |\beta|.$$

The division property is clear as soon as one realises that the multiples $\mu\beta$ of β lie at the corners of a lattice of squares in the complex plane. A typical square is the one with corners 0, β, $i\beta$, $(1 + i)\beta$. The remainder ρ is the difference between α and the nearest corner $\mu\beta$ in the lattice, so $0 \leq |\rho| < |\beta|$ because the distance $|\rho|$ between any point in a square and the nearest corner is less than the length $|\beta|$ of a side.

To find $\gcd(\alpha, \beta)$ one can therefore let $\alpha = \alpha_1$, $\beta = \beta_1$ and repeatedly compute

$$\alpha_{j+1} = \beta_j,$$
$$\beta_{j+1} = \text{remainder when } \alpha_j \text{ is divided by } \beta_j,$$

until a zero remainder β_k is obtained. (We use $j, k, \ldots$ for indices from now on, since i will be reserved for $\sqrt{-1}$.) Then $\gcd(\alpha, \beta) = \beta_k$, and there are Gaussian integers μ and ν such that $\gcd(\alpha, \beta) = \mu\alpha + \nu\beta$ by an argument like that used for $\mathbb{Z}$ (0.2.2). In fact, the rest of the route to unique prime factorisation is essentially the same as in $\mathbb{Z}$. We get the prime divisor property by arguing that

$$1 = \gcd(\pi, \alpha) = \mu\pi + \nu\alpha$$

when π is a Gaussian prime not dividing α, and multiplying both sides by β. Unique prime factorisation (up to unit factors) is obtained by supposing

$$\pi_1 \pi_2 \cdots \pi_r = \phi_1 \phi_2 \cdots \phi_s$$

are two prime factorisations of the same number, and cancelling $\pi_1, \pi_2, \ldots$ in turn until only units remain.

0.4.3 Gaussian primes

Identifying the actual primes of $\mathbb{Z}[i]$ is a separate question, and Gauss observed that it is closely connected with Fermat's two squares theorem. This is because $N(x + yi) = x^2 + y^2$, so divisors of $x + yi$ have norm dividing $x^2 + y^2$. In particular, if $x^2 + y^2$ is an ordinary prime then $x + yi$ is a Gaussian prime, because it cannot be the product of numbers of smaller norm. Fermat's two squares theorem tells us that the primes of the form $x^2 + y^2$ are exactly the primes $p = 4n + 1$ and $p = 2$, so each such $x^2 + y^2$ gives us two Gaussian primes, $x + yi$ and $x - yi$. Moreover, the factorisation

$$x^2 + y^2 = (x + yi)(x - yi)$$

is unique up to unit factors, by unique prime factorisation in $\mathbb{Z}[i]$. Hence there are exactly two Gaussian primes, up to unit factors, for each ordinary prime $p = 4n + 1$. (This shows, incidentally, that the partition $x^2 + y^2$ of p into two squares is unique, a result also stated by Fermat.) The ordinary prime 2 is exceptional, being the square of the Gaussian prime $1 + i$, up to a unit factor.

Conversely, if $x + yi$ is a Gaussian prime then so is its conjugate $x - yi$, since any factorisation of $x - yi$ yields a factorisation of $x + yi$ by conjugation. Thus

$$x^2 + y^2 = (x + yi)(x - yi)$$

is a Gaussian prime factorisation, and hence unique up to unit factors.

It follows that $x^2 + y^2$ is an ordinary prime, since any factorisation of it into ordinary primes would yield a different factorisation into Gaussian primes.

Thus the "properly complex" Gaussian primes are (up to unit factors) the conjugate factors $x + yi$ and $x - yi$ of ordinary primes of the form $x^2 + y^2$. By Fermat's two squares theorem these are the primes $p = 4n+1$ and $p = 2$. The real Gaussian primes are the ordinary primes $p = 4n+3$, since these have no divisors of the form $x^2 + y^2$ (by Fermat's theorem again) and hence no complex Gaussian prime divisors $x + yi$ (since a divisor $x + yi$ implies a divisor $x - yi$, by conjugation).

The close relation between Gaussian primes and the ordinary primes of the form $x^2 + y^2$ suggests that the theory of $\mathbb{Z}[i]$ may be used to prove Fermat's two squares theorem. Dedekind gave at least two such proofs, one in the memoir below (§27) and another in his final version of ideal theory (Dedekind (1894), §159). The latter proof is short enough to describe here, since it builds on the result of Lagrange that $p = 4n + 1$ divides a number of the form $m^2 + 1$, which we have already described in 0.3.4.

Since p does *not* divide either of the Gaussian factors $m + i$, $m - i$ of $m^2 + 1$ (the quotients $\frac{m}{p} \pm \frac{i}{p}$ are not Gaussian integers), p is not a Gaussian prime, by the prime divisor property of $\mathbb{Z}[i]$. We therefore have a Gaussian factorisation

$$p = (x + yi)(x - yi),$$

which gives

$$p = x^2 + y^2.$$

This argument from unique prime factorisation replaces Lagrange's argument from the equivalence of quadratic forms with discriminant -4. In fact, Dedekind (1894), §159, goes on to deduce the equivalence of these forms directly from unique prime factorisation in $\mathbb{Z}[i]$.

0.4.4 Imaginary quadratic integers

As an obvious generalisation of the Gaussian integers we can consider, for any positive integer c,

$$\mathbb{Z}[\sqrt{-c}] = \{x + y\sqrt{-c} : x, y \in \mathbb{Z}\},$$

with norm defined by

$$N(x + y\sqrt{-c}) = \left|x + y\sqrt{-c}\right|^2 = x^2 + cy^2.$$

The multiplicative property of the norm,

$$N((x + y\sqrt{-c})(x' + y'\sqrt{-c})) = N(x + y\sqrt{-c})N(x' + y'\sqrt{-c}),$$

is equivalent to the Brahmagupta identity (0.3.6) and one proves, as in $\mathbb{Z}[i]$, that each element of $\mathbb{Z}[\sqrt{-c}]$ has a factorisation into primes. However, the uniqueness of the prime factorisation depends on the value of c, as we shall see in the next section.

After the Gaussian integers ($c = 1$), the case $c = 2$ is also of interest, since $\mathbb{Z}[\sqrt{-2}]$ contains the numbers $x \pm \sqrt{-2}$ used by Euler in his solution of the equation $y^3 = x^2 + 2$. Unique prime factorisation also holds in $\mathbb{Z}[\sqrt{-2}]$. The proof, like that in $\mathbb{Z}[i]$, is based on the division property: *for any α and $\beta \neq 0$ there are μ and ρ in $\mathbb{Z}[\sqrt{-2}]$ ("quotient" and "remainder") such that*

$$\alpha = \mu\beta + \rho \quad where \quad 0 \leq |\rho| < |\beta|.$$

The division property holds in $\mathbb{Z}[\sqrt{-2}]$ for similar geometric reasons. The multiples $\mu\beta$ of β lie at the corners of a lattice of rectangles in the complex plane. A typical rectangle is the one with corners 0, β, $\sqrt{-2}\beta$, $(1 + \sqrt{-2})\beta$. The lengths of its sides are $|\beta|$ and $\sqrt{2}|\beta|$, so the maximum distance of any point from a corner (attained by the centre point) is $\sqrt{\frac{3}{4}}|\beta| < |\beta|$, as required for the division property.

It follows that $\mathbb{Z}[\sqrt{-2}]$ has a Euclidean algorithm, and hence unique prime factorisation, by the argument used for $\mathbb{Z}[i]$. Since $N(x + y\sqrt{-2}) = x^2 + 2y^2$, the only units in $\mathbb{Z}[\sqrt{-2}]$ are ± 1, and prime factorisation is unique up to sign (as in $\mathbb{Z}$, and hence "more unique" than in $\mathbb{Z}[i]$). The actual primes in $\mathbb{Z}[\sqrt{-2}]$ can be described using Fermat's theorem (0.3.2) that the (ordinary) primes of the form $x^2 + 2y^2$ are those of the form $p = 8n + 1$ or $p = 8n + 3$. Each ordinary prime $p = 8n + 1 = x^2 + 2y^2$ or $p = 8n + 3 = x^2 + 2y^2$ splits into primes $x + y\sqrt{-2}$ and $x - y\sqrt{-2}$ of $\mathbb{Z}[\sqrt{-2}]$, and each ordinary prime $p = 8n + 5$ or $p = 8n + 7$ remains prime in $\mathbb{Z}[\sqrt{-2}]$.

Finally, to justify Euler's solution of $y^3 = x^2 + 2$ it remains to prove that

$$\gcd(x + \sqrt{-2}, x - \sqrt{-2}) = 1$$

and that relatively prime numbers in $\mathbb{Z}[\sqrt{-2}]$ are cubes when their product is a cube.

To see why $\gcd(x + \sqrt{-2}, x - \sqrt{-2}) = 1$, note that any common divisor of $x + \sqrt{-2}$ and $x - \sqrt{-2}$ is also a divisor of their difference $2\sqrt{-2}$. Also, any solution x of $y^3 = x^2 + 2$ must be odd, because an even x yields an even y, in which case 8 divides y^3 and does *not* divide $x^2 + 2$. Thus $N(x + \sqrt{-2})$ must be odd, and hence its gcd with $N(2\sqrt{-2}) = 8$ is 1. But this implies that 1 is the gcd of $x - \sqrt{-2}$ and $2\sqrt{-2}$, and hence of $x + \sqrt{-2}$ and $x - \sqrt{-2}$.

Thanks to unique prime factorisation, and the fact that the units of $\mathbb{Z}[\sqrt{-2}]$ are ± 1, the proof that relatively prime numbers are cubes when their product is a cube is essentially the same as in $\mathbb{Z}$. This in turn is like the proof for squares mentioned in 0.2.1.

0.4.5 The failure of unique prime factorisation

It should not be a great surprise that something goes wrong in $\mathbb{Z}[\sqrt{-5}]$, because the norm of $x + y\sqrt{-5}$ is the anomalous quadratic form $x^2 + 5y^2$. In short, prime factorisation is not unique. Consider the following two factorisations of 6:

$$2 \times 3 = (1 + \sqrt{-5})(1 - \sqrt{-5}).$$

The norms of the factors are $N(2) = 4$, $N(3) = 9$, $N(1 + \sqrt{-5}) = 6$ and $N(1 - \sqrt{-5}) = 6$. The only nontrivial divisors of these norms are 2 and 3, which are not norms, hence the factors 2, 3, $(1 + \sqrt{-5})$, $(1 - \sqrt{-5})$ themselves have no nontrivial divisors. These factors are therefore primes in $\mathbb{Z}[\sqrt{-5}]$, and they certainly differ by more than unit factors, since the norms of the factors on the left are different from the norms of the factors on the right.

So here we have another anomalous behaviour of $x^2 + 5y^2$. Its irrational factors $x + y\sqrt{-5}$ and $x - y\sqrt{-5}$ belong to a set $\mathbb{Z}[\sqrt{-5}]$ of "integers" without unique prime factorisation. Can this be related to the fact that $x^2 + 5y^2$ is one of two inequivalent quadratic forms with discriminant -20?

Indeed yes. The nonuniqueness of prime factorisation in $\mathbb{Z}[\sqrt{-5}]$ can be traced to the existence of ordinary integers $x^2 + 5y^2$ with divisors not of the same form. If each number $x + y\sqrt{-5}$ had a unique prime factorisation in $\mathbb{Z}[\sqrt{-5}]$,

$$x + y\sqrt{-5} = (x_1 + y_1\sqrt{-5})(x_2 + y_2\sqrt{-5}) \cdots (x_r + y_r\sqrt{-5}),$$

then we should have the following factorisation of its norm in $\mathbb{Z}$:

$$x^2 + 5y^2 = (x_1^2 + 5y_1^2)(x_2^2 + 5y_2^2) \cdots (x_r^2 + 5y_r^2).$$

And each factor $(x_j^2 + 5y_j^2)$ would be an ordinary prime or the square of an ordinary prime, since any further decomposition of $x^2 + 5y^2$ into ordinary primes would yield a different prime factorisation of $x + y\sqrt{-5}$ in $\mathbb{Z}[\sqrt{-5}]$. Then, since the product of numbers of the form $(x_j^2 + 5y_j^2)$ is another number of the same form (by Brahmagupta's identity), each divisor of $x^2 + 5y^2$ would also have this form. But of course 6 has divisors 2 and 3, which are *not* of the form $x^2 + 5y^2$. Instead, they are of the *other* form with discriminant -20, namely $2x^2 + 2xy + 3y^2$.

Most of this could have been noticed by Lagrange, but the failure of unique prime factorisation in $\mathbb{Z}[\sqrt{-5}]$ was not explicitly pointed out until Kummer (1844) proposed the introduction of "ideal factors" to save it. Kummer (1846b) went on to define a notion of equivalence for ideal factors, under which ideal factor classes correspond to classes of forms with the same discriminant. Thus inequivalence of forms is actually the same thing as failure of unique prime factorisation. Kummer's discovery changed the whole direction of number theory at this point, from the theory of forms to the theory of ideal factors. However, it would be an oversimplification to say that this happened just because of quadratic forms like $x^2 + 5y^2$. A second major influence was the theory of roots of unity, which will be discussed in the next section.

It is possible that number theory could have changed direction much earlier, if not for the conservatism of Gauss. According to Kummer (1846a), Gauss was aware of the failure of unique prime factorisation for quadratic integers when he wrote the *Disquisitiones*, and could see that something like ideal factorisation was needed. However, he was unable to find a rigorous description of ideal factors, and invented the theory of composition of forms as a substitute. He later told Dirichlet

If I wanted to proceed with the use of imaginaries in the way that earlier mathematicians have done, then one of my earlier researches which is very difficult [composition of forms] could have been done in a very simple way. (Excerpt from Kummer (1846a) in Edwards (1977), p. 335.)

Having invested so much energy in composition of forms, Gauss perhaps became unwilling to pursue the alternative theory of quadratic integers. As we know, it was only in 1832 (the paper on Gaussian integers) that he got as far as proving unique prime factorisation for $\mathbb{Z}[i]$.

On the other hand, it must be admitted that Gauss's work on other topics – the theory of regular polygons (or circle division) and the search for reciprocity laws – was crucial in the development of the more general concept of algebraic integer. As we shall now see, this work also belongs to the theory of roots of unity.

0.5 Roots of unity

0.5.1 Fermat's last theorem

The Pythagorean equation is probably the most fruitful equation in number theory, having been the inspiration for important results of Euclid (0.2.1), Diophantus (0.2.4) and Fermat (0.3.1). Diophantus threw new light on the Pythagorean equation by considering rational solutions, instead of just integer solutions, and by showing that *any* nonzero square in $\mathbb{Q}$ splits into two nonzero squares. Next to this result in Diophantus' *Arithmetica*, Fermat wrote his famous marginal note:

It is impossible to separate a cube into two cubes, or a biquadrate into two biquadrates, or in general any power higher than second into powers of like degree; I have discovered a truly marvellous proof of this which however this margin is too small to contain.

This claim of Fermat's became known as *Fermat's last theorem* in the early 19th century, as by then it was the only claim of Fermat's still to be proved or disproved.

Fermat's last theorem is the claim that the equation

$$a^n + b^n = c^n$$

has no nonzero solution $a, b, c \in \mathbb{Q}$ when n is an integer > 2. Since any rational solution becomes an integer solution when multiplied through by a common denominator, an equivalent claim is that $a^n + b^n = c^n$ has no nonzero *integer* solution. Although these two claims are logically equivalent, they suggest different viewpoints and different approaches to proving the the theorem. The approach through rational numbers, which stems from Diophantus, is "geometric" in a broad sense. The approach through integers, which stems from Fermat and Euler, is "algebraic". The geometric approach seems to have won the day, but only by calling on large amounts of algebra, analysis and topology as well. The algebraic approach, while falling short of Fermat's last theorem, was an outstanding success in the development of algebra as a whole. In particular, it was the main stimulus for the development of Dedekind's theory of ideals.

Fermat's own contribution to Fermat's last theorem was slight, as far as we know. The "marvellous proof" he claimed in his marginal note was probably based on a mistake, as he did not repeat the claim later, and left only the proof for $n = 4$ (0.3.1). This proof does not generalise to other values of n, and it is the *only* case of Fermat's last theorem that has been proved by elementary methods.

The search for an algebraic proof of Fermat's last theorem really begins
with the *Algebra* of Euler (1770). As we have seen in 0.4.1, this book
contains a brilliant, and basically sound, treatment of the equation $y^3 = x^2 + 2$. It also contains a treatment of $x^3 + y^3 = z^3$, the Fermat equation
for $n = 3$. The reasoning is not quite so sound, but it is extremely
thought-provoking, and pregnant with possibilities for generalisation.

Euler rewrites the equation as

$$y^3 = z^3 - x^3,$$

and then factorises the right hand side into

$$(z - x)(z - \zeta_3 x)(z - \zeta_3^2 x)$$

where

$$\zeta_3 = \frac{-1 + \sqrt{-3}}{2}, \quad \zeta_3^2 = \frac{-1 - \sqrt{-3}}{2}$$

are the imaginary cube roots of 1. At this point Euler makes an inter-
esting mistake. He wants to argue that the factors $z - x$, $z - \zeta_3 x$, $z - \zeta_3^2 x$
are relatively prime factors of a cube, hence cubes themselves, but uses
this argument in the wrong setting – in $\mathbb{Z}[\sqrt{-3}]$ rather than $\mathbb{Z}[\zeta_3]$. It so
happens that unique prime factorisation *fails* in $\mathbb{Z}[\sqrt{-3}]$, as can be seen
from the example

$$4 = 2 \times 2 = (1 + \sqrt{-3})(1 - \sqrt{-3}),$$

whose factors all have minimal norm, 4, for nonunit elements of $\mathbb{Z}[\sqrt{-3}]$.
However, unique prime factorisation is *valid* in $\mathbb{Z}[\zeta_3]$, for geometric reasons like those that apply to $\mathbb{Z}[i]$ and $\mathbb{Z}[\sqrt{-2}]$, and Euler's idea can be
made to work.

0.5.2 The cyclotomic integers

The numbers $\mathbb{Z}[\zeta_3]$ in the proof of Fermat's last theorem for $n = 3$ are
examples of *cyclotomic integers*. The general ring of cyclotomic integers
is $\mathbb{Z}[\zeta_n]$ where

$$\zeta_n = \cos \frac{2\pi}{n} + i \sin \frac{2\pi}{n}$$

is a complex n^{th} root of 1. It was first studied by Gauss (1801), for
geometric reasons. The points $\zeta_n, \zeta_n^2, \ldots, \zeta_n^n = 1$ are equally spaced

around the unit circle, hence the name "cyclotomic", from the Greek for "circle-dividing". Since

$$\zeta_n^n - 1 = (\zeta_n - 1)(\zeta_n^{n-1} + \cdots + \zeta_n + 1) \quad \text{and} \quad \zeta_n \neq 1$$

it follows that ζ_n satisfies the equation

$$\zeta_n^{n-1} + \cdots + \zeta_n + 1 = 0,$$

and hence $\mathbb{Z}[\zeta_n]$ can be described explicitly as

$$\mathbb{Z}[\zeta_n] = \{a_0 + a_1\zeta_n + \cdots + a_{n-2}\zeta_n^{n-2} : a_0, a_1, \ldots, a_{n-2} \in \mathbb{Z}\}.$$

When n is a prime, $z^{n-1} + \cdots + z + 1 = 0$ is called the *cyclotomic equation*. Gauss (1801), article 341, proved that it is the equation of minimal degree, with integer coefficients, satisfied by ζ_n. When n is not prime the cyclotomic equation is of degree lower than $n - 1$, and is not so easy to describe.

Gauss was initially interested in $\mathbb{Z}[\zeta_n]$ because of its bearing on the construction of the regular n-gon by straightedge and compass, which is possible when ζ_n is expressible in terms of rational operations and square roots. He made the amazing discovery that this occurs for precisely the n that are products of a power of 2 with distinct primes of the form $2^{2^h} + 1$. The only known primes of this form are 3, 5, 17, 257, 65537, but the latter three yield regular n-gons not previously constructed. In fact, Gauss's construction of the regular 17-gon was the first such construction since ancient times, and was an important factor in his decision to become a mathematician.

Perhaps the most surprising feature of this discovery is that primes $2^{2^h} + 1$ had already been considered – for an entirely different reason – by Fermat. Fermat mistakenly believed the numbers $2^{2^h} + 1$ to be prime for *all* values of h (Fermat (1640a)). This is false. Euler found that 641 divides $2^{2^5} + 1$, and since then no more such primes have been found. Yet in a way Fermat was not far wrong – what *is* true is that no two numbers $2^{2^h} + 1$ have a common prime factor. This was observed by Pólya and Szegö (1924), and used by them to give a new proof that there are infinitely many primes. The numbers $2^{2^h} + 1$ are now called *Fermat numbers*, and the primes among them are called *Fermat primes*.

Gauss's discovery may have closed a chapter in the history of geometry, but it opened a new one in the history of number theory. Quite apart from drawing attention to the problem of finding all Fermat primes (intractable so far), the theory of cyclotomic integers created new links between different parts of number theory. In particular, Gauss found connections between arithmetic modulo p, the quadratic integers $\mathbb{Z}[\sqrt{\pm p}]$ and the cyclotomic integers $\mathbb{Z}[\zeta_p]$. As we shall explain in the next section, the connection between cyclotomic integers and quadratic integers came to light because a solution of the cyclotomic equation by square roots is needed to construct the regular p-gon.

Cyclotomic integers also arise naturally in connection with Fermat's last theorem, because of the factorisation

$$z^n - x^n = (z - x)(z - \zeta_n x) \cdots (z - \zeta_n^{n-1} x),$$

which generalises the factorisation used by Euler in his attempt to prove Fermat's last theorem for $n = 3$. In fact, Lamé (1847) used this formula in an attempt to prove Fermat's last theorem for arbitrary $n > 2$. He assumed, as would follow from unique prime factorisation, that relatively prime factors of an nth power are themselves n^{th} powers. Unfortunately, he failed to check uniqueness of prime factorisation in $\mathbb{Z}[\zeta_n]$, and this omission turned out to be fatal. Unique prime factorisation is crucial, and it *fails* in $\mathbb{Z}[\zeta_{23}]$. Unbeknownst to Lamé, Kummer (1844) had already discovered this, and had taken steps to deal with the problem. However, this is getting ahead of our story. Let us return to Gauss.

0.5.3 Cyclotomic integers and quadratic integers

Since a straightedge creates lines (which have equations of degree 1) and the compass creates circles (which have equations of degree 2), constructions find intersections of lines and circles, hence they solve linear and quadratic equations, and this can be done by rational operations and square roots. Thus if the regular p-gon is constructible by straightedge and compass, there is necessarily a solution of the cyclotomic equation

$$z^{p-1} + \cdots + z + 1 = 0$$

by rational operations and square roots. The roots are $\zeta_p, \zeta_p^2, \ldots, \zeta_p^{p-1}$, hence the equation says that the sum of all the roots is -1. Gauss solves the equation by stepwise evaluation of sums of progressively smaller subsets of the roots, now known as *Gauss sums*. In the case $p - 1 = 2^{2^h}$, each subset is half the size of the previous one, and its sum is obtainable from the preceding Gauss sum by square roots. Thus in 2^h steps one reaches a single root of the cyclotomic equation, expressible by square roots alone.

For example a root ζ of the cyclotomic equation for $p = 5$ satisfies

$$\zeta^4 + \zeta^3 + \zeta^2 + \zeta + 1 = 0,$$

which can rewritten

$$(\zeta^3 + \zeta^2)^2 + (\zeta^3 + \zeta^2) - 1 = 0,$$

since $(\zeta^3 + \zeta^2)^2 = \zeta^4 + 2\zeta^5 + \zeta$ and $\zeta^5 = 1$. Thus we have a quadratic equation for $\zeta^3 + \zeta^2$, which can be solved by square roots, and we can then find ζ^3 and ζ^2 individually by solving quadratic equations, since we know their sum and product.

Of course, if $p - 1$ is not a power of 2 one cannot repeatedly halve the number of terms in the Gauss sums until only one term remains. However, in the nontrivial case where p is odd, the first halving is always possible. From now on we shall use ζ, without subscript, to denote an arbitrary root of the cyclotomic equation. The set of all roots ζ^k is partitioned into those for which k is a square mod p, and those for which k is a nonsquare. As we know from the existence of primitive roots (0.3.5), these sets are of equal size. In fact, we know that the squares mod p are the even powers of a primitive root mod p, and the nonsquares are its odd powers. In the case $p = 5$, 2 is a primitive root, so its odd powers are 2 and $2^3 = 8 \equiv 3 \pmod 5$, and the corresponding Gauss sum is the one used above: $\zeta^3 + \zeta^2$. We could also have used its complement, $\zeta^4 + \zeta$, whose exponents are the squares mod 5.

In the *Disquisitiones*, article 356, Gauss showed that the sums whose exponents are the squares and nonsquares mod p are the two roots of the equation

$$x^2 + x \pm \frac{p \pm 1}{4} = 0,$$

with the $-$ sign when $p \equiv 1 \pmod 4$ and the $+$ sign when $p \equiv 3 \pmod 4$. Since the roots are $-\frac{1}{2} \pm \frac{\sqrt{p}}{2}$ in the former case and $-\frac{1}{2} \pm i\frac{\sqrt{p}}{2}$ in the latter, it follows that, if a is a primitive root mod p,

$$\sum_j \zeta^{a^{2j}} - \sum_j \zeta^{a^{2j+1}} = \pm\sqrt{p} \quad \text{when} \quad p \equiv 1 \pmod 4,$$

$$\sum_j \zeta^{a^{2j}} - \sum_j \zeta^{a^{2j+1}} = \pm i\sqrt{p} \quad \text{when} \quad p \equiv 3 \pmod 4.$$

The left hand sides of these equations can be written more concisely with the help of the *Legendre symbol* or *quadratic character symbol* defined by

$$\left(\frac{k}{p}\right) = \begin{cases} +1 & \text{if } k \text{ is a square mod } p \\ -1 & \text{if } k \text{ is a nonsquare mod } p. \end{cases}$$

Quadratic characters have the following *multiplicative property*,

$$\left(\frac{k}{p}\right)\left(\frac{l}{p}\right) = \left(\frac{kl}{p}\right),$$

which follows immediately from the fact that the squares mod p are precisely the even powers of a primitive root.

In terms of quadratic characters, the Gauss sum on the left of the equations is simply $S = \sum_{k=1}^{p-1} \left(\frac{k}{p}\right)\zeta^k$, and Gauss's theorem is equivalent to:

$$S^2 = \begin{cases} +p & \text{if } p \equiv 1 \pmod 4 \\ -p & \text{if } p \equiv 3 \pmod 4. \end{cases}$$

Even more concisely,

$$S^2 = \left(\frac{-1}{p}\right)p$$

since we know from 0.3.5 that -1 is a square mod p if and only if $p \equiv 1 \pmod 4$.

The proof that $S^2 = \left(\frac{-1}{p}\right)p$ goes as follows. Since

$$S = \sum_{k=1}^{p-1} \left(\frac{k}{p}\right)\zeta^k,$$

$$S^2 = \sum_{k,l=1}^{p-1} \left(\frac{k}{p}\right)\left(\frac{l}{p}\right)\zeta^{k+l}$$

$$= \sum_{k,l=1}^{p-1} \left(\frac{kl}{p}\right)\zeta^{k+l}$$

by the multiplicative property of quadratic characters.

Now each $k \not\equiv 0 \pmod p$ has a multiplicative inverse mod p (0.3.1), hence as k runs through the nonzero congruence classes mod p, so does kl, for fixed $l \not\equiv 0 \pmod p$. We may therefore replace k by kl, and

$$S^2 = \sum_{k,l=1}^{p-1} \left(\frac{kl^2}{p}\right)\zeta^{kl+l}$$

$$= \sum_{k,l=1}^{p-1} \left(\frac{k}{p}\right)\zeta^{l(k+1)}$$

since k is a square mod $p \Leftrightarrow kl^2$ is a square mod p

$$= \sum_{l=1}^{p-1} \left(\frac{p-1}{p}\right) \zeta^{lp} + \sum_{k=1}^{p-2} \left(\frac{k}{p}\right) \left(\sum_{l=1}^{p-1} \zeta^{l(k+1)}\right)$$

$$= \sum_{l=1}^{p-1} \left(\frac{p-1}{p}\right) + \sum_{k=1}^{p-2} \left(\frac{k}{p}\right) (\zeta + \zeta^2 + \cdots + \zeta^{p-1})$$

since $l(k+1)$ for fixed $k < p-1$ runs through the nonzero congruence classes

$$= \left(\frac{p-1}{p}\right) (p-1) + \sum_{k=1}^{p-2} \left(\frac{k}{p}\right) (-1)$$

by the cyclotomic equation

$$= \left(\frac{p-1}{p}\right) (p-1) + \sum_{k=1}^{p-1} \left(\frac{k}{p}\right) (-1) - \left(\frac{p-1}{p}\right)$$

$$= \left(\frac{p-1}{p}\right) p$$

since half the $\left(\frac{k}{p}\right)$ are $+1$ and half are -1, by 0.3.5

$$= \left(\frac{-1}{p}\right) p \quad \text{since} \quad p-1 \equiv -1 \pmod{p}.$$

It follows that $S = \pm\sqrt{(\frac{-1}{p})p}$, and hence $\mathbb{Z}[\zeta_p]$ contains either $\sqrt{p}$ or $\sqrt{-p}$. Thus Gauss's computation forges links between squares mod p, $\mathbb{Z}[\zeta_p]$ and $\mathbb{Z}[\sqrt{\mp p}]$. In the next section we shall how Gauss exploited these links to give a proof of quadratic reciprocity.

Remark. The relationship of cyclotomic integers with quadratic integers is special, and there is no comparable relationship with, say, cubic integers. For example, $\sqrt[3]{2}$ does not belong to any $\mathbb{Z}[\zeta_p]$. One way to see this is to extend $\mathbb{Z}[\zeta_p]$ to the field $\mathbb{Q}(\zeta_p)$, by allowing arbitrary quotients, and to consider automorphisms of fields. Automorphisms of $\mathbb{Q}(\zeta_p)$ extend functions of the form $\tau_i(\zeta_p) = \zeta_p^i$, and hence any two of them commute. On the other hand, any field containing $\sqrt[3]{2}$ and ζ_3 has noncommuting automorphisms, extending all permutations of the set $\{\sqrt[3]{2}, \zeta_3 \sqrt[3]{2}, \zeta_3^2 \sqrt[3]{2}\}$. We shall not elaborate on these hints, since the ideas belong to Galois theory, which Dedekind wishes to avoid. However, they do show that Galois theory is just over the horizon.

0.5.4 Quadratic reciprocity

The *quadratic reciprocity law* is the following symmetric relationship between the quadratic characters of two odd primes p, q.

$$\left(\frac{p}{q}\right)\left(\frac{q}{p}\right) = \begin{cases} -1 & \text{if } p,q \equiv 3 \ (\text{mod } 4) \\ +1 & \text{otherwise.} \end{cases}$$

This surprising relationship can be extracted from the qth power of the Gauss sum $S = \sum_{k=1}^{p-1} \left(\frac{k}{p}\right)\zeta^k$ by using the "mod q binomial theorem"

$$(x+y)^q \equiv x^q + y^q \pmod{q}.$$

As mentioned in 0.3.1, this holds for q prime because in that case all the binomial coefficients $\binom{q}{1}, \ldots, \binom{q}{q-1}$ are divisible by q. This leads to the "mod q multinomial theorem"

$$(x_1 + x_2 + \cdots + x_k)^q \equiv x_1^q + x_2^q + \cdots + x_k^q \pmod{q}$$

by induction on k, when $x_1, x_2, \ldots, x_k$ are indeterminates. We are able to treat the powers $\zeta, \zeta^2, \ldots, \zeta^{p-1}$ as indeterminates because the minimality of the cyclotomic polynomial (0.5.2) means that each element of $\mathbb{Z}[\zeta]$ is uniquely expressible as a sum $\sum_{k=1}^{p-1} a_k \zeta^k$. Here is the qth power calculation:

$$S^q = \left(\sum_{k=1}^{p-1}\left(\frac{k}{p}\right)\zeta^k\right)^q$$

$$\equiv \sum_{k=1}^{p-1}\left(\frac{k}{p}\right)^q \zeta^{kq} \pmod{q}$$

$$= \sum_{k=1}^{p-1}\left(\frac{k}{p}\right)\zeta^{kq} \quad \text{since } \left(\frac{k}{p}\right)^n = \left(\frac{k}{p}\right) \text{ for any odd power } n$$

$$= \sum_{k=1}^{p-1}\left(\frac{q}{p}\right)^2\left(\frac{k}{p}\right)\zeta^{kq} \quad \text{since } \left(\frac{q}{p}\right)^2 = 1$$

$$= \left(\frac{q}{p}\right)\sum_{k=1}^{p-1}\left(\frac{kq}{p}\right)\zeta^{kq} \quad \text{by the multiplicative property}$$

$$= \left(\frac{q}{p}\right)S$$

since, for fixed q, kq runs through the nonzero congruence classes. We now multiply both sides by S and use $S^2 = \left(\frac{-1}{p}\right) S$, obtaining

$$S^{q+1} \equiv \left(\frac{q}{p}\right)\left(\frac{-1}{p}\right) S \pmod{q}.$$

The left hand side is

$$(S^2)^{\frac{q+1}{2}} = \left(\frac{-1}{p}\right)^{\frac{q+1}{2}} p^{\frac{q+1}{2}}$$

and hence we can cancel $\left(\frac{-1}{p}\right) p$ from both sides to obtain

$$\left(\frac{-1}{p}\right)^{\frac{q-1}{2}} p^{\frac{q-1}{2}} \equiv \left(\frac{q}{p}\right) \pmod{q}.$$

Finally, Euler's criterion (0.3.5) allows us to replace $p^{\frac{q-1}{2}}$ by the quadratic character $\left(\frac{p}{q}\right)$, and since $\left(\frac{-1}{p}\right) = \pm 1$ we have

$$\left(\frac{p}{q}\right) = \left(\frac{q}{p}\right) \quad \text{if} \quad q \equiv 1 \pmod 4,$$

$$\left(\frac{-1}{p}\right)\left(\frac{p}{q}\right) = \left(\frac{q}{p}\right) \quad \text{if} \quad q \equiv 3 \pmod 4,$$

from which the statement of quadratic reciprocity follows, using the value of $\left(\frac{-1}{p}\right)$, the quadratic character of -1, known from 0.3.5.

Remarks. The law of quadratic reciprocity has probably been proved in more ways than any other theorem in mathematics except the theorem of Pythagoras. Gauss himself gave eight proofs, of which the above is the sixth (Gauss (1818)). There are shorter or more elementary proofs, but the one using Gauss sums is especially enlightening, because of the way it relates squares mod p to quadratic integers and cyclotomic integers. It was certainly an inspiration to Dedekind, and his own proof (§27 of the memoir) shows how Gauss's ideas can be transformed with the help of ideals. Dedekind says he is giving the sixth proof of Gauss, but he succeeds in eliminating all the computations. This is exactly what he intends to accomplish with the theory of ideals. As he says in §12:

It is preferable, as in the modern theory of functions, to seek proofs based immediately on fundamental characteristics, rather than on calculation, and indeed to construct the theory in such a way that it is able to predict the results of calculation.

0.5.5 Other reciprocity laws

The appearance of roots of unity in the proof of quadratic reciprocity
is very interesting and historically important, for the reasons just mentioned, but it is not strictly necessary. There are many other proofs
involving just arithmetic mod p and q. One reason it is easy to avoid
complex numbers is that the real numbers $+1$ and -1 can serve as values
of the quadratic character $\left(\frac{k}{q}\right)$.

If, on the other hand, one wants to define a multiplicative "cubic
character" $\left(\frac{k}{q}\right)_3$, with three different values according as k is of the
form a^{3n}, a^{3n+1} or a^{3n+2} (where a is a primitive root mod p), then real
values of the character will not work. In fact, the obvious values are

$$\left(\frac{k}{q}\right)_3 = \begin{cases} 1 & \text{if } k = a^{3n} \\ \zeta_3 & \text{if } k = a^{3n+1} \\ \zeta_3^2 & \text{if } k = a^{3n+2}. \end{cases}$$

Similarly, the obvious values of the "biquadratic character" $\left(\frac{k}{q}\right)_4$ are
$1, i, -1$ and $-i$. Thus there is a more pressing reason to use roots of
unity when one seeks reciprocity laws for 3rd and 4th powers. The latter
was in fact what Gauss was doing when he developed the theory of
unique prime factorisation in $\mathbb{Z}[i]$. Likewise, Eisenstein (1844) found a
cubic reciprocity law through investigation of $\mathbb{Z}[\zeta_3]$.

Kummer even declared that cyclotomic integers were more important
for their application to higher reciprocity laws than their application to
Fermat's last theorem. In 1847 he said:

The Fermat theorem is indeed more of a curiosity than a main point in the
science [of number theory]. (Kummer (1847b).)

And a few years later, when he had made progress on reciprocity laws:

Through my investigations in the theory of complex numbers [cyclotomic in-
tegers] and their applications to the proof of the Fermat theorem, which I
have communicated to the Academy of Sciences over the past three years, I
have succeeded in discovering the general reciprocity law for arbitrary power
residues, which in the present state of number theory stands as the main
problem and pinnacle of the discipline. (Kummer (1850b).)

However, he said this when he was still flushed with success over reciprocity – and when his work on Fermat's last theorem was running out
of steam. In the long run not everyone agreed with him. When Hilbert
surveyed algebraic number theory in his influential *Zahlbericht* (Hilbert
(1897)), he gave pride of place to Kummer's results on Fermat's last

theorem. Even today, when Kummer's approach to Fermat's last theorem has been abandoned, its ideas are at the core of algebraic number theory.

0.6 Algebraic integers

0.6.1 Definition

The examples of quadratic and cyclotomic integers hint at a general concept of algebraic integer, but it is not clear how it should be defined. In fact, it is not even clear how *quadratic* integers should be defined in general. An example which underlines the subtlety of the concept is

$$\mathbb{Z}[\sqrt{-3}] = \{m + n\sqrt{-3} : m, n \in \mathbb{Z}\}.$$

As already mentioned in 0.5.1, unique prime factorisation fails in $\mathbb{Z}[\sqrt{-3}]$, but it is valid in $\mathbb{Z}[\zeta_3]$, where $\zeta_3 = \frac{-1+\sqrt{-3}}{2}$. Thus we would be happier if all members of $\mathbb{Z}[\zeta_3]$ were classed as integers, and not just the members of $\mathbb{Z}[\sqrt{-3}]$. At the same time, of course, we do not want to admit all the members of

$$\mathbb{Q}[\sqrt{-3}] = \{s + t\sqrt{-3} : s, t \in \mathbb{Q}\}.$$

This would certainly be going too far, since all rational numbers would be included.

A definition which draws the line at just the right place is the one given by Dedekind in the introduction to his memoir. He states it in a way that makes it a natural specialisation of the concept of algebraic number:

A number θ is called an *algebraic* number if it satisfies an equation

$$\theta^n + a_1\theta^{n-1} + a_2\theta^{n-2} + \cdots + a_{n-1}\theta + a_n = 0$$

with finite degree n and rational coefficients $a_1, a_2, \ldots, a_{n-1}, a_n$. It is called an *algebraic integer*, or simply an *integer*, when it satisfies an equation of the form above in which all the coefficients $a_1, a_2, \ldots, a_{n-1}, a_n$ are rational integers.

This definition has some desirable properties, as we shall see in the next section, but it also has one undesirable property: if α is an algebraic integer then so is $\sqrt{\alpha}$, hence every algebraic integer has a nontrivial factorisation $\alpha = \sqrt{\alpha}\sqrt{\alpha}$. This is very different from the behaviour of ordinary integers, and the theory of *all* algebraic integers is therefore not very useful in the study of $\mathbb{Z}$, for the reasons given in 0.1.

Dedekind escaped this situation by restricting attention to the integers in a *field of finite degree*, which can be most simply defined as the closure

$\mathbb{Q}(\alpha)$ of $\mathbb{Q} \cup \{\alpha\}$ under rational operations, where α is an algebraic number. The *integers of* $\mathbb{Q}(\alpha)$ are simply the algebraic integers in $\mathbb{Q}(\alpha)$. When α is an algebraic integer they include all members of $\mathbb{Z}[\alpha]$ but sometimes more. For example, the set of integers in $\mathbb{Q}(\sqrt{-3})$ is not $\mathbb{Z}[\sqrt{-3}]$ but $\mathbb{Z}[\zeta_3]$. On the other hand, the set of integers of $\mathbb{Q}(\zeta_p)$ is indeed $\mathbb{Z}[\zeta_p]$. Kummer, working without a general definition of algebraic integer, actually defined his integers to be the members of $\mathbb{Z}[\zeta_p]$, so in a sense he was lucky to hit on the "correct" integers.

0.6.2 Basic properties

Dedekind first stated his definition of algebraic integer in his Supplement X, §160, to the second edition of Dirichlet's *Vorlesungen über Zahlentheorie* (Dedekind (1871)). There he added that:

It follows immediately that a rational number is an algebraic integer if and only if it is a rational integer.

This result goes back at least as far as Gauss. In article 11 of the *Disquisitiones* he assumes the reader knows that a rational solution of a monic polynomial equation with integer coefficients is itself an integer.

It follows that an ordinary integer a does not divide an ordinary integer b, as an algebraic integer, unless a divides b as an ordinary integer. This is obviously crucial when results about ordinary integers are being derived as special cases of results about algebraic integers. Similarly crucial properties of algebraic integers are closure under $+$, $-$ and $\times$. These follow from another property of monic polynomial equations, first pointed out by Eisenstein (1850).

If $f(x) = 0$ is a monic polynomial equation with ordinary integer coefficients and roots $\alpha_1, \alpha_2, \ldots, \alpha_n$, and if $g(\alpha_1, \ldots \alpha_n)$ is any polynomial in the roots with ordinary integer coefficients, then $g(\alpha_1, \ldots \alpha_n)$ is also a root of a monic polynomial equation $h(x) = 0$ with ordinary integer coefficients.

Eisenstein's proof uses the theorem of Newton (1665) that symmetric polynomials in the roots are polynomial functions of the coefficients, applying it to the equation

$$h(x) = \prod_{\substack{\text{all permutations } \sigma \text{ of } 1,2,\ldots,n}} (x - g(\alpha_{\sigma(1)}, \ldots, \alpha_{\sigma(n)})) = 0.$$

Obviously $h(x)$ is monic, with symmetric coefficients, hence it actually has integer coefficients, and its roots include $g(\alpha_1, \ldots, \alpha_n)$. Taking

$g(\alpha_1, \alpha_2) = \alpha_1 + \alpha_2$, $\alpha_1 - \alpha_2$ or $\alpha_1\alpha_2$ shows that $\alpha_1 + \alpha_2$, $\alpha_1 - \alpha_2$, $\alpha_1\alpha_2$ all satisfy monic equations with ordinary integer coefficients, and hence they are algebraic integers.

Dedekind remarks in §17 of the memoir that he wishes to avoid the theory of symmetric functions. In fact in §13 (and previously in §160 of his Supplement X to Dirichlet (1871)) he has already derived closure from the fact that a system of homogeneous linear equations with nonzero solution must have nonzero determinant. These proofs, along with the concepts of basis and independence he also introduced, were important in the development of linear algebra as a self-contained discipline. Setting the theory of algebraic integers within the theory of fields of finite degree gave this development an extra push, because these fields are vector spaces of finite dimension over the rationals. The reason is simple: if the minimal rational polynomial equation satisfied by α is of degree n, then $1, \alpha, \alpha^2, \ldots, \alpha^{n-1}$ is a basis for $\mathbb{Q}(\alpha)$ over $\mathbb{Q}$.

Fields of finite degree are important to the concept of algebraic integer because they admit a suitable concept of norm. Again generalising the situation for quadratic and cyclotomic integers, Dedekind defines the *norm* $N(\omega)$ of any number $\omega \in \mathbb{Q}(\alpha)$. If $\alpha, \alpha_1, \alpha_2, \ldots, \alpha_{n-1}$ are the roots of the irreducible monic equation satisfied by α, and $\phi(\alpha)$ is the rational function of α equal to ω, then $N(\omega)$ is the product of $\phi(\alpha), \phi(\alpha_1), \phi(\alpha_2), \ldots, \phi(\alpha_{n-1})$, the so-called *conjugates* of ω. As Dedekind explains in §16 and §17 of the memoir, the conjugates of ω are also its images under all the isomorphisms of $\mathbb{Q}(\alpha)$ onto other number fields. The latter fact makes it easy to see the *multiplicative* property of the norm: $N(\omega_1\omega_2) = N(\omega_1)N(\omega_2)$. It turns out that $N(\omega)$ is always a rational number, and $N(\omega)$ is a rational integer when ω is an integer of $\mathbb{Q}(\alpha)$. This means that every integer has a divisor of minimal norm, and hence a factorisation into "primes" (though the factorisation is not necessarily unique).

0.6.3 Class numbers

As already mentioned in 0.3.3, Lagrange (1773) gave the first rigorous results bearing on unique prime factorisation of algebraic integers. His reduction process for quadratic forms gives an easy way to find the class number of quadratic forms with negative discriminant, because the reduced forms are inequivalent, apart from some simple exceptions. The class number of $x^2 - cy^2$ can be redefined and reinterpreted as the class number of the *field* $\mathbb{Q}(\sqrt{c})$ (and generalised to any field of finite degree, see §29 of the monograph). Lagrange's computations were extended by Gauss (1801), who showed that the class number of $\mathbb{Q}(\sqrt{c})$ is 1 for $c = -1, -2, -3, -7, -11, -19, -43, -67, -163$ (*Disquisitiones*, article 303). He also conjectured that these are the *only* negative c for which $\mathbb{Q}(\sqrt{c})$ has class number 1, a conjecture which remained open until 1966, when it was proved by Baker and Stark. (See Baker (1966) and Stark (1967).) An equivalent statement is that these are the only quadratic fields with negative discriminant and unique prime factorisation.

The situation is more difficult for quadratic fields with positive discriminant. There is a similar reduction process, also due to Lagrange, which shows there are only finitely many reduced forms, but it is no longer clear which of them are equivalent. Thus we do not immediately know the class number, only that it is finite. Gauss made some progress, giving an algorithm in article 195 of the *Disquisitiones* to decide equivalence of reduced forms. Dirichlet (1839) made even greater progress, using analysis to discover a *formula* for the class number of quadratic fields. While analytic methods are beyond our scope, Dirichlet's result does indicate the depth of the problem. No more elementary approach has yet been found, and his related theorem on primes in arithmetic progressions (Dirichlet (1837)) has likewise not been proved in a more elementary manner.

Apparently, Dirichlet also found an analytic formula for the class number of the cyclotomic field $\mathbb{Q}(\zeta_p)$. Kummer (1847a) mentions this, while giving a direct proof that the class number of $\mathbb{Q}(\zeta_p)$ is finite. He gives the following specific values for which the class number is 1: $p = 5, 7, 11, 13, 17, 19$. Thus unique prime factorisation holds in these fields, and the first failure is in $\mathbb{Q}(\zeta_{23})$, which Kummer shows to have class number 3. Dirichlet never published his class number formula for $\mathbb{Q}(\zeta_p)$, presumably deferring to Kummer, and Kummer (1850a) gave his own version.

The finiteness of the class number for all fields of finite degree was first proved by Dedekind (1871), in his first account of ideal theory. The proof is similar to the one given in the memoir below.

0.6.4 Ideal numbers and ideals

The long story of unique prime factorisation, lost and regained, can now be summarised as follows. Euclid discovered unique prime factorisation in $\mathbb{Z}$, to all intents and purposes, when he showed that a prime divides a product only if it divides one of the factors. Fermat saw trouble coming in the behaviour of $x^2 + 5y^2$, Lagrange found what the trouble was (class number greater than 1), and Gauss found an elaborate way round it (composition of forms). However, Gauss's way out was a retreat from the direction already opened by Lagrange – the use of algebraic integers in the study of $\mathbb{Z}$.

The first to embrace the algebraic integers, for all their faults, was Kummer. He could see that unique prime factorisation was lost, but he hoped to regain it:

It is greatly to be lamented that this virtue of the real numbers [i.e. the rational integers], to be decomposable into prime factors, always the same ones for a given number, does not also belong to the complex numbers [i.e. the integers of cyclotomic fields]; were this the case, the whole theory, which is still laboring under such difficulties, could easily be brought to its conclusion. For this reason, the complex numbers we have been considering seem imperfect, and one may well ask whether one ought not to look for another kind which would preserve the analogy with the real numbers with respect to such a fundamental property. (Translation by Weil (1975) from Kummer (1844).)

The first sentence is the one quoted by Dedekind in Latin in his introduction: "Maxime dolendum videtur ..." (Kummer's paper was one of the last important mathematical works written in Latin.) Dedekind goes on to say:

But the more hopeless one feels about the prospects of later research on such numerical domains, the more one has to admire the steadfast efforts of Kummer, which were finally rewarded by a truly great and fruitful discovery.

This was the discovery of ideal numbers. The rest of the story can be left to Dedekind himself. He explains Kummer's ideal numbers with admirable clarity, and his own distillation of the concept of ideal from them. He also gives a candid account of the difficulties en route to unique prime factorisation of prime ideals. The concept of ideal is very simple, and so are the concepts of divisor and product, but unfortunately it is *not* clear that if ideal $\mathfrak{b}$ divides ideal $\mathfrak{a}$ then $\mathfrak{a} = \mathfrak{b}\mathfrak{c}$ for some ideal $\mathfrak{c}$. Dedekind was stymied by this difficulty for a long time, and here he explains how he overcame it after several years of struggle. The memoir is not only a superb exposition, but also a rare opportunity to see a great mathematician wrestling with a problem, with blow by blow comments up to the final victory.

0.7 The reception of ideal theory

0.7.1 How the memoir came to be written

Dedekind's first exposition of ideal theory, Dedekind (1871), included proofs of unique prime ideal factorisation and finiteness of the class number, together with a very impressive application of the theory: a correspondence between the quadratic forms of discriminant D and the ideals of $\mathbb{Q}(\sqrt{D})$, under which the product of ideals corresponds to composition of forms. Replacement of the complicated and mysterious forms by objects that behaved like integers should have been a revelation, but Dedekind's contemporaries were slow to appreciate his achievement.

There was great resistance to the idea of treating infinite sets as mathematical objects. Dedekind had tried this in 1857 when he introduced congruence *classes* in place of specific residues, and again in 1872 with his definition of real numbers as Dedekind sections. But he was fighting 2000 years of tradition (plus a formidable modern opponent, Leopold Kronecker). The "horror of infinity" that had haunted mathematics since Zeno was not to be dispelled overnight. Most mathematicians were not even willing to consider a theory based on infinite sets, let alone appreciate its power or elegance.

Nevertheless, Dedekind tried again. The memoir on algebraic integers first appeared in instalments in the *Bulletin des sciences mathématiques et astronomiques* in 1877. The reason Dedekind chose this unusual outlet was a letter he received from Rudolf Lipschitz in 1876:

Bonn, 11 March 1876

Dear Colleague,

To make the purpose of these lines clear, it is necessary to mention a few things first. You have perhaps seen that Darboux' Bulletin has brought out a series of my works on the theory of homogeneous functions of an arbitrary number of differentials. Herr Darboux invited me to write such a series in 1872, leaving the language up to me. The second editor of the Bulletin, Herr Hoüel, who also translated Dirichlet's German works for Liouville's Journal, made a careful French translation, and since Herr Hoüel has always been very attentive and obliging, I have remained in touch with him since. As I recently had cause to write to him, I brought up an idea I have long cherished, namely concerning your own investigations in the second edition of Dirichlet's lectures on number theory, from §159 onward. In my opinion they are of rare value, and in Germany itself they have not received their full due. Therefore, it would be highly desirable if you could arrange to give a detailed analysis of these investigations in the Bulletin. Herr Hoüel has now sent a reply, which I can best convey to you in his own words, and that is the purpose of my letter. He writes: [French text follows, a translation of which is] Our bulletin would receive with great pleasure a detailed analysis of M. Dedekind's work.

I have communicated the substance of your letter to M. Darboux, who will himself write to M. Dedekind, and who will in any case be happy to publish his work. If you have occasion to write to the learned professor, you may tell him how flattered we will be by his collaboration, and thank him in advance for anything he is disposed to send us. [End of French text.]

Of course, I cannot know whether you are inclined to undertake such a work. But I hope you will understand the true motive for the steps I have taken, namely, the keen desire for the mathematical public to learn the true value of first rate research.

With cordiality and deep respect,

Yours,

R. Lipschitz

(Translation of Lipschitz (1986), pp. 47–48.)

On April 29 Dedekind wrote a long reply, the first part of which is:

Dear Colleague,

Your letter brought me great and unexpected joy, since for years I had more or less given up hope of interesting anybody in my general theory of ideals. With the exception of Professor H. Weber in Königsberg, with whom I have worked closely as editor of the forthcoming collected works of Riemann, and who has expressed his intention to acquaint himself with this theory, you are the first, not merely to show interest in the subject, but also in such a practical way, that it revives hope that my work may not have been in vain. I thought that the inclusion of this investigation in Dirichlet's *Zahlentheorie* would be the best way to attract a wider circle of mathematicians to the field, but little by little I have become convinced that the presentation itself is to blame for the failure of this plan. I can only suppose that the presentation deterred readers through excessive brevity and terseness, and since autumn I have been spending my free time, obtained by resigning my three year directorship of the local polytechnic, working out a more detailed presentation of the theory of ideals, and I have come so far as to obtain a somewhat improved form of the essential foundation (the content of §163). (Translation of Dedekind's letter to Lipschitz, Lipschitz (1986), pp. 48–49.)

Correspondence about the series of articles continued until 12 August 1876, as the details were worked out and Hoüel was engaged to translate Dedekind's manuscript into French. After their publication in the *Bulletin*, the articles were also published as a book (Dedekind (1877)).

0.7.2 Later development of ideal theory

Dedekind continued to present new versions of his ideal theory, revising it substantially for the third (1879) and fourth (1894) editions of Dirichlet's *Zahlentheorie*, and expanding its interaction with Galois theory. He and Weber also made a major application of ideal theory outside number theory, with their arithmetic theory of Riemann surfaces (Dedekind and Weber (1882)). This is the great fruit of their collaboration in the publication of Riemann's works – not only a turning point in the development of complex function theory, but also the beginning of modern algebraic geometry. (The "modern theory of functions" cited by Dedekind in §12 of the memoir as the inspiration for proofs "based on fundamental characteristics, rather than on calculation," is undoubtedly another reference to Riemann's work.) Dedekind and Weber developed an ideal theory for functions, with polynomials playing the role of "rational" integers, and entire algebraic functions playing the role of "algebraic" integers. Once again, prime ideals are important, and unique prime factorisation is valid (in fact, somewhat easier to prove).

In the same year, Kronecker (1882) published a long account of his own theory of fields and integers, which he had been developing for some decades. His theory included an equivalent approach to unique prime factorisation, though expressed in a very different, and less readable, language. Kronecker was almost diametrically opposite to Dedekind in mathematical philosophy; he opposed nonconstructive methods, infinity, and in particular the use of infinite sets as mathematical objects. He even claimed not to believe in irrational numbers. Dedekind favoured the free use of infinite sets wherever it was necessary or convenient. His definitions of irrational numbers (as pairs of sets of rationals) and "ideal numbers" (as the sets he called ideals) should have made the convenience of his position clear, but in fact neither the Kronecker nor the Dedekind approach attracted an immediate following.

As Artin (1962) said:

Dedekind's presentation is easy to read and elegant for us today, but at the time it was too modern. Thus the appearance of Hilbert's *Zahlbericht* in the *Jahresbericht der Deutschen Mathematikervereinigung* in 1897 was greeted with great joy. Hilbert presented all the results known up to that time and, through great simplifications, made the results of Kummer accessible to a wider circle of readers. (Translation of Artin (1962), p. 549.)

The *Zahlbericht*, as its name suggests, was intended to be a report on number theory. Hilbert used groups, rings, fields and modules heavily, but with the clear aim of elucidating properties of numbers. In particular, he used ideal theory to simplify Kummer's results on the Fermat problem.

It was only in the 1920s that Artin and Emmy Noether created modern ideal theory by abstracting the properties of rings that make unique factorisation into prime ideals possible, and using these properties as axioms. The subject then became part of ring theory, and Dedekind's theories of algebraic numbers and algebraic functions became mere special cases, at least in theory. In practice, one still finds the best of both worlds in the theory of algebraic numbers.

Acknowledgements

I am indebted to George Francis for sending me a copy of the Lipschitz–Dedekind correspondence when it was not available in Australia. Thanks also to Peter Stevenhagen and John McCleary for corrections and comments, and to two anonymous reviewers at Cambridge University Press for valuable advice.

Clayton, Victoria, Australia   John Stillwell

Bibliography

Artin, E. (1962). Die Bedeutung Hilberts für die moderne Mathematik. *Jahrbuch Akad. Wiss. Göttingen.* Also in his *Collected Papers*, 547–551.

Baker, A. (1966). Linear forms in the logarithms of algebraic numbers. *Mathematika*, **13**, 204–216.

Colebrooke, H. T. (1817). *Algebra, with Arithmetic and Mensuration, from the Sanscrit of Brahmegupta and Bháscara.* John Murray, London. Reprinted by Martin Sandig, Wiesbaden, 1973.

Cox, D. A. (1989). *Primes of the Form $x^2 + ny^2$.* Wiley.

Dedekind, R. (1871). Supplement X. In Dirichlet (1871), 380–497.

Dedekind, R. (1872). *Stetigkeit und Irrationalzahlen.* English translation *Continuity and Irrational Numbers*, in *Essays on the Theory of Numbers*, Open Court, Chicago, 1901.

Dedekind, R. (1877). *Sur la Théorie des Nombres Entiers Algébriques.* Gauthier-Villars.

Dedekind, R. (1894). Supplement XI. In Dirichlet (1894), 434–657.

Dedekind, R. and Weber, H. (1882). Theorie der algebraischen Functionen einer Veränderlichen. *J. reine und angew. Math.*, **92**, 181–290.

Dirichlet, P. G. L. (1837). Beweis der Satz, dass jede unbegrentze arithmetische Progression, deren erstes Glied und Differenz ganze Zahlen ohne gemeinschaftliche Factor sind, unendliche viele Primzahlen enthält. *Abh. der Königl. Preuss. Akad. Wiss.*, 45–81. Also in his *Mathematische Werke*, volume 1, 313–342.

Dirichlet, P. G. L. (1839). Recherches sur diverses applications de l'analyse infinitésimal à la théorie des nombres. *J. reine angew. Math.*, **19**, 324–369. Also in his *Mathematische Werke*, volume 1, 411–496.

Dirichlet, P. G. L. (1871). *Vorlesungen über Zahlentheorie.* Vieweg, second edition.

Dirichlet, P. G. L. (1894). *Vorlesungen über Zahlentheorie.* Vieweg, fourth edition. Reprinted by Chelsea, 1968.

Edwards, H. M. (1977). *Fermat's Last Theorem.* Springer-Verlag.

Eisenstein, F. G. (1844). Beweis des Reciprocitätssatzes für die cubischen Reste in der Theorie der aus dritten Wurzeln der Einheit zusammengesetzten complexen Zahlen. *J. reine angew. Math.*, **27**, 289–310. Also in his *Mathematische Werke*, volume 1, 59–80.

Eisenstein, F. G. (1850). Über einige allgemeine Eigenschaften der Gleichung, von welcher die Theorie der ganzen Lemniscate abhängt, nebst Anwendungen derselben auf die Zahlentheorie. *J. reine angew. Math.*, **39**, 224–287. Also in his *Mathematische Werke*, volume 2, 556–619.

Euler, L. (1744). Theoremata circa divisores numerorum in hac forma *paa ± qbb* contentorum. *Comm. acad. sci. Petrop.*, **14**, 151–181. Also in his *Opera Omnia* ser. I, volume 2, 194–222.

Euler, L. (1749). Letter to Goldbach, 12 April 1749. In Euler (1843), 493–495.

Euler, L. (1756). Solutio generalis quorundam problematum diophanteorum quae vulgo nonnisi solutiones speciales admittere videntur. *Novi comm. acad. sci. Petrop.*, **6**, 155–184. Also in his *Opera Omnia* ser. I, volume 2, 428–445.

Euler, L. (1761). Theoremata arithmetica nova methodo demonstrata. *Novi comm. acad. sci. Petrop.*, **8**, 74–104. In his *Opera Omnia*, ser. I, volume 2, 531–555.

Euler, L. (1770). *Vollständige Einleitung zur Algebra*. English translation *Elements of Algebra*, Springer-Verlag, 1984.

Euler, L. (1843). *Correspondance Mathématique et Physique.* St. Petersburg. Reprinted by Johnson Reprint Corporation, 1968.

Fermat, P. (1640a). Letter to Frenicle, 18 October 1640. In Fermat (1894), 209.

Fermat, P. (1640b). Letter to Frenicle, August(?) 1640. In Fermat (1894), 205–206.

Fermat, P. (1640c). Letter to Mersenne, 25 December 1640. In Fermat (1894), 212.

Fermat, P. (1654). Letter to Pascal, 25 September 1654. In Fermat (1894), 310–314.

Fermat, P. (1657). Letter to Digby, 15 August 1657. In Fermat (1894), 345.

Fermat, P. (1670). Observations sur Diophant. In Fermat (1896), 241–276.

Fermat, P. (1894). *Œuvres*, volume 2. Gauthier-Villars.

Fermat, P. (1896). *Œuvres*, volume 3. Gauthier-Villars.

Gauss, C. F. (1801). *Disquisitiones Arithmeticae*. English translation, Yale University Press, 1966.

Gauss, C. F. (1818). Theorematis fundamentalis in doctrina de residuis quadraticis demonstrationes et ampliationes novae. *Comm. Soc. Reg. Sci. Gött. Rec.*, **4**. Also in his *Werke*, volume 2, 49–64.

Gauss, C. F. (1832). Theoria residuorum biquadraticorum. *Comm. Soc. Reg. Sci. Gött. Rec.*, **7**. Also in his *Werke*, volume 2, 67–148.

Heath, T. L. (1910). *Diophantus of Alexandria*. Cambridge University Press.

Heath, T. L. (1925). *The Thirteen Books of Euclid's Elements*. Cambridge University Press, second edition. Reprinted by Dover, 1956.

Hilbert, D. (1897). Die Theorie der algebraischen Zahlkörper. *Jber. deutsch. Math. Verein.*, **4**, 175–546. Also in his *Gesammelte Abhandlungen*, volume 1, 63–363.

Kronecker, L. (1870). Auseinandersetzung einige Eigenschaften der Klassenzahl idealer complexer Zahlen. *Monatsber. Königl. Akad. Wiss. Berlin*, 881–889. Also in his *Werke* I, 271–282.

Kronecker, L. (1882). Grundzüge einer Theorie der algebraischen Grössen. *J. reine und angew. Math.*, **92**, 1–122.

Kummer, E. E. (1844). De numeris complexis, qui radicibus unitatis et numeris realibus constant. *Gratulationschrift der Univ. Breslau zur*

Jubelfeier der Univ. Königsberg. Also in his *Collected Papers*, volume 1, 165–192.

Kummer, E. E. (1846a). Letter to Kronecker, 14 June 1846. In Kummer (1975), 98.

Kummer, E. E. (1846b). Zur Theorie der complexen Zahlen. *Monatsber. Akad. Wiss. Berlin*, 87–96. Also in his *Collected Papers*, volume 1, 203–210.

Kummer, E. E. (1847a). Beweis der Fermat'schen Satzes der Unmöglichkeit von $x^\lambda + y^\lambda = z^\lambda$ für eine unendliche Anzahl Primzahlen λ. *Monatsber. Akad. Wiss. Berlin*, 132–141, 305–319. Also in his *Collected Papers*, volume 1, 274–297.

Kummer, E. E. (1847b). Über die Zerlegung der aus Wurzeln der Einheit gebildeten complexen Zahlen in ihre Primfactoren. *J. reine angew. Math.*, 327–367. Also in his *Collected Papers*, volume 1, 211–251.

Kummer, E. E. (1850a). Allgemeine Reciprocitätsgesetze für beliebig hohe Potenzreste. *Monatsber. Akad. Wiss. Berlin*, 154–165. Also in his *Collected Papers*, volume 1, 345–357.

Kummer, E. E. (1850b). Bestimmung der Anzahl nicht äquivalenter Classen für die aus λten Wurzeln der Einheit gebildeten complexen Zahlen und die idealen Factoren derselben. *J. reine angew. Math.*, 93–116. Also in his *Collected Papers*, volume 1, 299–322.

Kummer, E. E. (1975). *Collected Papers*. Springer-Verlag.

Lagrange, J. L. (1768). Solution d'un problème d'arithmétique. *Miscellanea Taurinensia*, **4**, 19ff. Also in his *Œuvres*, volume 1, 671–731.

Lagrange, J. L. (1770). Nouvelle méthode pour résoudre les problèmes indéterminés en nombres entiers. *Mém. de l'acad. roy. sci. Berlin*, **24**. Also in his *Œuvres*, volume 2, 655–726.

Lagrange, J. L. (1773). Recherches d'arithmétique. *Nouv. mém. de l'acad. sci. Berlin*, 265ff. Also in his *Œuvres*, volume 3, 695–795.

Lamé, G. (1847). Démonstration général du théorème de Fermat. *Comptes Rendus*, **24**, 310–315.

Legendre, A.-M. (1798). *Essai sur la Théorie des Nombres.* Duprat, Paris. Third edition, entitled *Théorie des Nombres* (1830), reprinted by Blanchard, 1955.

Lipschitz, R. (1986). *Briefwechsel mit Cantor, Dedekind, Helmholtz, Kronecker, Weierstraß und anderen.* Deutsche Mathematiker Vereinigung.

Neugebauer, O. and Sachs, A. (1945). *Mathematical Cuneiform Texts.* Yale University Press.

Newton, I. (1665). Of the nature of equations. In Newton (1967), 519–520.

Newton, I. (1967). *The Mathematical Papers of Isaac Newton*, volume 1. Cambridge Universtity Press.

Pólya, G. and Szegö, G. (1924). *Aufgaben und Lehrsätze aus der Analysis.* Springer-Verlag. English translation *Problems and Theorems in Analysis*, Springer-Verlag, 1976.

Stark, H. M. (1967). A complete determination of the complex quadratic fields of class-number one. *Michigan Math. J.*, **14**, 1–27.

Weil, A. (1974). Two lectures on number theory, past and present. *Enseign. Math.*, **20**, 87–110. Also in his *Collected Papers*, volume 3, 279–302.

Weil, A. (1975). Introduction to Kummer (1975).

Weil, A. (1984). *Number Theory: an Approach Through History.* Birkhäuser.

Part two

Theory of algebraic integers

Introduction

In response to the invitation it has been my honour to receive, I propose, in the present memoir, to develop the *fundamental principles* of the general theory of algebraic integers I published in the second edition of Dirichlet's *Vorlesungen über Zahlentheorie*. Because of the extraordinary scope of this field of mathematical research, however, I restrict myself here to the pursuit of a single goal, which I shall try to clarify in the remarks that follow.

The theory of divisibility of numbers, which is the basis of arithmetic, was established in its essentials by Euclid. At any rate, the fundamental theorem that each integer is uniquely decomposable into a product of primes is an immediate consequence of the theorem, proved by Euclid,[†] that a product of two numbers is not divisible by a prime unless the prime divides one of the factors.

Two thousand years later, Gauss gave, for the first time, an extension of the notion of integer. The numbers $0, \pm 1, \pm 2, \ldots$ previously going under that name, and which I shall call *rational integers* from now on, were extended when Gauss introduced[‡] *complex integers* of the form $a + b\sqrt{-1}$, where a and b are any rational integers. He showed that the general laws of divisibility of these numbers are identical with those that regulate the domain of rational integers.

The broadest generalisation of the notion of integer is the following. A number θ is called an *algebraic* number if it satisfies an equation

$$\theta^n + a_1\theta^{n-1} + a_2\theta^{n-2} + \cdots + a_{n-1}\theta + a_n = 0,$$

with finite degree n and rational coefficients $a_1, a_2, \ldots, a_{n-1}, a_n$. It is

---

† *Elements*, VII, 32.
‡ *Theoria residuorum biquadraticorum*, II; 1832.

called an *algebraic integer*, or simply an *integer*, when it satisfies an equation of the form above in which all the coefficients $a_1, a_2, \ldots, a_{n-1}, a_n$
are rational integers. It follows immediately from this definition that the
sum, difference and product of integers are also integers. Consequently,
an integer α will be said to be *divisible* by an integer β if $\alpha = \beta\gamma$, where
γ is likewise an integer. An integer ϵ will be called a *unit* when every
integer is divisible by ϵ. By analogy, a *prime* must be an integer α which
is not a unit and which is divisible only by units ϵ and products of the
form $\epsilon\alpha$. However, it is easy to see that, in the domain of all integers
we are considering at present, primes do not exist, since every integer
which is not a unit is always the product of two, or rather any number,
of integral factors which are not units.

Nevertheless, the existence of primes and the analogy with the domains of rational or complex integers re-emerges when we restrict ourselves to an infinitely small part of the domain of all integers, in the
following manner. If θ is an algebraic number there is, among the infinitely many equations with rational coefficients satisfied by θ, exactly
one

$$\theta^n + a_1\theta^{n-1} + \cdots + a_{n-1}\theta + a_n = 0,$$

of minimal degree, and which we call for this reason *irreducible*. If x_0,
$x_1, x_2, \ldots, x_{n-1}$ denote arbitrary rational numbers, the numbers of the
form

$$\phi(\theta) = x_0 + x_1\theta + x_2\theta^2 + \cdots + x_{n-1}\theta^{n-1},$$

the set of which we call Ω, will also be algebraic numbers, and they enjoy
the fundamental property that their sums, differences, products and
quotients also belong to the set Ω. I call such a set Ω a *field of degree*† n.
The numbers $\phi(\theta)$ belonging to a field Ω are now partitioned, following
the definition above, into two large sets: the set $\mathfrak{o}$ of integers and the
nonintegral, or fractional, numbers. *The problem we set ourselves is to
establish the general laws of division that govern such a system* $\mathfrak{o}$.

The system $\mathfrak{o}$ is evidently identical with the system of all rational
integers when $n = 1$, or with the complex integers when $n = 2$ and
$\theta = \sqrt{-1}$. Certain phenomena which occur in these two special domains
$\mathfrak{o}$ occur again in every domain $\mathfrak{o}$ of this nature. Above all, the unlimited
decomposition which prevails in the domain of all algebraic integers is
never encountered in a domain $\mathfrak{o}$ of the kind indicated, as one easily
sees by consideration of norms. If we define the *norm* of any number

† Dedekind calls it "finite, of degree n". (Translator's note.)

$\mu = \phi(\theta)$ belonging to the field Ω to be the product

$$N(\mu) = \mu\mu_1\mu_2\ldots\mu_{n-1},$$

whose factors are the conjugate numbers

$$\mu = \phi(\theta), \quad \mu_1 = \phi(\theta_1), \quad \mu_2 = \phi(\theta_2), \quad \ldots, \quad \mu_{n-1} = \phi(\theta_{n-1})$$

where $\theta, \theta_1, \theta_2, \ldots, \theta_{n-1}$ denote all the roots of the irreducible n^{th} degree equation, then $N(\mu)$ will always be, as we know, a rational number, and never 0 unless $\mu = 0$. At the same time, one always has

$$N(\alpha\beta) = N(\alpha)N(\beta)$$

where α and β are any two numbers of the field Ω. Now if μ is an integer, and hence a number in $\mathfrak{o}$, then the conjugate numbers $\mu_1, \mu_2, \ldots, \mu_{n-1}$ will likewise be integers, and so $N(\mu)$ will be a rational integer.

This norm plays an extremely important role in the theory of numbers in the domain $\mathfrak{o}$. In fact, let any two numbers α, β in this domain be called *congruent* or *incongruent* relative to the third μ, called the *modulus*, according as their difference $\pm(\alpha - \beta)$ is or is not divisible by μ. Then we can, just as in the theory of rational or complex integers, partition all numbers of the system $\mathfrak{o}$ into *number classes*, such that each class is the set of all numbers congruent to a given number (representing the class). And a deeper study shows us that the number of classes (with the exception of the case $\mu = 0$) is always finite, equal at most to the absolute value of $N(\mu)$. An immediate consequence of this result is that $N(\mu) = \pm 1$ just in case μ is a unit. Now if a number in the system $\mathfrak{o}$ is called *decomposable* when it is the product of two numbers of the system, neither of which is a unit, it evidently follows from the above that each decomposable number can always be represented as the product of a finite number of *indecomposable* factors.

This result again corresponds completely with the law holding in the theory of rational or complex integers, namely that each composite number is representable as the product of a finite number of prime factors. But at the same time this is the point where the analogy, observed until now with the old theory, is in danger of being irrevocably broken. In his researches on the domain of numbers belonging to the theory of circle division, hence corresponding to equations of the form $\theta^m = 1$, Kummer noted the existence of a phenomenon distinguishing the numbers of that domain from those considered previously. They differ in a manner so complete and so essential as to leave little hope of preserving the simple laws that govern the old theory of numbers. In fact, whereas each number in the domain of rational or complex integers decomposes *uniquely* into a product of primes, one discovers that, in the numerical domains considered by Kummer, a number may be representable *in several entirely different ways* as the product of indecomposable numbers. Or what amounts to the same thing, one discovers that the *indecomposable* numbers lack the characteristic of genuine *primes*, inasmuch as a prime cannot divide a product of two or more factors without dividing at least one of the factors.

But the more hopeless one feels about the prospects of later research on such numerical domains,† the more one has to admire the steadfast efforts of Kummer, which were finally rewarded by a truly great and fruitful discovery. That geometer succeeded in resolving all the apparent irregularities in the laws.‡ By considering the indecomposable numbers which lack the characteristics of true primes to be products of *ideal* prime factors whose effect is only apparent when they are combined together, he obtained the surprising result that the laws of divisibility in the numerical domains studied by him were now in complete agreement with those that govern the domain of rational integers. Each number which is not a unit behaves consistently in all divisibility situations, whether as divisor or dividend, as a prime or as a product of prime factors, actual or ideal. Two ideal numbers, whether prime or composite, which yield two actual numbers when combined with the same ideal number are called *equivalent*, and all the ideal numbers equivalent to the same ideal number form a *class of ideal numbers*. The set of all actual numbers, considered as a special case of ideal numbers, forms the *principal class*. To each class there corresponds an infinite system of equivalent homogeneous *forms*, in n variables and of degree n, which are decomposable into n linear factors with algebraic coefficients. The number of these classes is finite, and Kummer succeeded in determining their number by extending the principles used by Dirichlet to determine the number of classes of binary quadratic forms.

The great success of Kummer's researches in the domain of circle division allows us to suppose that the same laws hold in *all* numerical

---

† In the memoir: *De numeris complexis qui radicibus unitatis et numeris integri realibus constant* (Vrastislaviæ, 1844, §8), Kummer said "Maxime dolendum videtur, quod hæc numerorum realium virtus, ut in factores primes dissolvi possint qui pro eodem numero semper iidem sint, non eadem est numerorum complexorum, quæ si esset tota hæc doctrina, quæ magnis adhuc difficultatibus laborat, facile absolvi et ad finem perduci posset". (See the Introduction, Section 0.6.4, for an English translation of this passage. Translator's note.)

‡ *Zur Theorie der complexen Zahlen* (*Crelle's Journal*, 35).

domains ο of the most general kind considered above. In my researches, the goal of which has been to arrive at a definitive answer to this question, I began by building on the theory of higher order congruences, since I had previously noticed that the latter theory allows Kummer's researches to be shortened considerably. However, while this method leads to a point very close to my goal, I have not been able to surmount, by this route, certain exceptions to the laws holding in other cases. I did not achieve the general theory, without exceptions, that I first published in the place mentioned above, until I abandoned the old formal approach and replaced it by another; a fundamentally simpler conception focussed directly on the goal. In the latter approach I need no concept more novel than that of Kummer's *ideal numbers*, and it is sufficient to consider a *system of actual numbers* that I call an *ideal*. The power of this concept resides in its extreme simplicity, and my plan being above all to inspire confidence in this notion, I shall try to explain the train of thought that led me to it.

Kummer did not define ideal numbers themselves, but only the divisibility of these numbers. If a number α has a certain property A, to the effect that α satisfies one or more congruences, he says that α is divisible by an ideal number corresponding to the property A. While this introduction of new numbers is entirely legitimate, it is nevertheless to be feared at first that the language which speaks of ideal numbers being determined by their products, presumably in analogy with the theory of rational numbers, may lead to hasty conclusions and incomplete proofs. And in fact this danger is not always completely avoided. On the other hand, a precise definition covering *all* the ideal numbers that may be introduced in a particular numerical domain ο, and at the same time a general definition of their multiplication, seems all the more necessary since the ideal numbers do not actually exist in the numerical domain ο. To satisfy these demands it will be necessary and sufficient to establish once and for all the common characteristic of the properties $A, B, C, \ldots$ that serve to introduce the ideal numbers, and then to indicate how one can derive, from properties A, B corresponding to particular ideal numbers, the property C corresponding to their product.†

† The legitimacy, or rather the necessity, of such demands, which must always be imposed with the introduction or creation of new arithmetic elements, becomes more evident when compared with the introduction of *real irrational* numbers, which was the subject of a pamphlet of mine (*Stetigkeit und irrationale Zahlen*, Brunswick, 1872). Assuming that the arithmetic of *rational* numbers is soundly based, the question is how one should introduce the irrational numbers and define the operations of addition, subtraction, multiplication and division on them. My

This problem is essentially simplified by the following considerations. Since a characteristic property A serves to define, not an ideal number itself, but only the divisibility of the numbers in o by the ideal number, one is naturally led to consider the set a of *all* numbers α of the domain o which are divisible by a particular ideal number. I now call such a system an *ideal* for short, so that for each particular ideal number there corresponds a particular ideal a. Now if, conversely, the property A of divisibility of a number α by an ideal number is equivalent to the membership of α in the corresponding ideal a, one can consider, in place of the properties $A, B, C, \ldots$ defining the ideal numbers, the corresponding ideals a, b, c, ..., in order to establish their common and exclusive character. Bearing in mind that these ideal numbers are introduced with no other goal than restoring the laws of divisibility in the numerical domain o to complete conformity with the theory of rational numbers, it is evidently necessary that the numbers actually existing in o, and which are always present as factors of composite numbers, be regarded

---

first demand is that arithmetic remain free from intermixture with extraneous elements, and for this reason I reject the definition of real number as the ratio of two quantities of the same kind. On the contrary, the definition or creation of irrational number ought to be based on phenomena one can already define clearly *in the domain R* of rational numbers. Secondly, one should demand that all real irrational numbers be engendered simultaneously by a common definition, and not successively as roots of equations, as logarithms, etc. Thirdly, the definition should be of a kind which also permits a perfectly clear definition of the calculations (addition, etc.) one needs to make on the new numbers. One achieves all of this in the following way, which I only sketch here:

1. By a *section* of the domain R I mean any partition of the rational numbers into two categories such that each number of the first category is algebraically less than every number of the second category.

2. Each particular rational number a *engenders* a particular section (or two sections, not essentially different) in which each rational number is in the first or second category according as it is smaller or larger than a (while a itself can be assigned at will to either category).

3. There are infinitely many sections which *cannot* be engendered by rational numbers in the manner just described. For each section of this kind one creates or introduces into arithmetic a special *irrational* number, corresponding to the section (or engendered by it).

4. If α, β are any two real numbers (rational or irrational), then one easily defines $\alpha > \beta$ or $\alpha < \beta$ in terms of the sections they engender. Moreover, one can easily define, in terms of these two sections, the four sections corresponding to the sum, difference, product and quotient of the two numbers α, β. In this way the four fundamental arithmetic operations are defined without any obscurity for an arbitrary pair of real numbers, and one can really *prove* propositions such as, for example, $\sqrt{2} \cdot \sqrt{3} = \sqrt{6}$, which had not previously been done, as far as I know, in the strict sense of the word.

5. When defined in this way, the irrational numbers unite with the rational numbers to form a domain $\mathfrak{R}$ without gaps and *continuous*. Each section of this domain $\mathfrak{R}$ is produced by a particular number of the domain itself; it is impossible to engender new numbers in this domain $\mathfrak{R}$.

as a special case of ideal numbers. Thus if μ is a particular number of $\mathfrak{o}$, the system $\mathfrak{a}$ of all numbers $\alpha = \mu\omega$ in the domain $\mathfrak{o}$ divisible by μ likewise has the essential character of an ideal, and it will be called a *principal ideal*. The latter system is evidently not altered when one replaces μ by $\epsilon\mu$, where ϵ is any unit in $\mathfrak{o}$. Now, the notion of integer established above immediately yields the following two elementary theorems on divisibility:

1. If two integers $\alpha = \mu\omega$, $\alpha' = \mu\omega'$ are divisible by the integer μ, then so are their sum $\alpha+\alpha' = \mu(\omega+\omega')$ and their difference $\alpha-\alpha' = \mu(\omega-\omega')$, since the sum $\omega + \omega'$ and difference $\omega - \omega'$ of two integers ω, ω' are themselves integers.

2. If $\alpha = \mu\omega$ is divisible by μ, each number $\alpha\omega' = \mu(\omega\omega')$ divisible by α will also be divisible by μ, since each product $\omega\omega'$ of integers ω, ω' is itself an integer.

If we apply these theorems, true for all integers, to the numbers ω of our numerical domain $\mathfrak{o}$, with μ denoting a particular one of these numbers and $\mathfrak{a}$ the corresponding principal ideal, we obtain the following two fundamental properties of such a numerical system $\mathfrak{a}$:

I. *The sum and difference of any two numbers in the system $\mathfrak{a}$ are always numbers in the same system $\mathfrak{a}$.*

II. *Any product of a number in the system $\mathfrak{a}$ by a number of the system $\mathfrak{o}$ is a number in the system $\mathfrak{a}$.*

Now, as we pursue the goal of restoring the laws of divisibility in the domain $\mathfrak{o}$ to complete conformity with those ruling the domain of rational integers, by introducing ideal numbers and a corresponding language, it is apparent that the definitions of these ideal numbers and their divisibility should be stated in such a way that the elementary theorems 1 and 2 above remain valid not only when the number μ is actual, but also when it is ideal. Consequently, the properties I and II should hold not only for principal ideals, but for *all* ideals. We have therefore found a common characteristic of all ideals: to each actual or ideal number there corresponds a unique ideal $\mathfrak{a}$, enjoying the properties I and II.

A fact of the highest importance, which I was able to prove rigorously only after numerous vain attempts, and after surmounting the greatest difficulties, is that, conversely, each system enjoying properties I and II is also an ideal. That is, it is the set $\mathfrak{a}$ of all numbers α of the domain $\mathfrak{o}$ divisible by a particular number; either an actual number or an ideal number indispensable for the completion of the theory. Properties I and II are therefore not just necessary, but also sufficient conditions for a numerical system $\mathfrak{a}$ to be an ideal. Any other condition imposed on the numerical system $\mathfrak{a}$, if it is not simply a consequence of properties I and II, makes a complete explanation of all the phenomena of divisibility in the domain $\mathfrak{o}$ impossible.

This finding naturally led me to base the theory of numbers in the domain $\mathfrak{o}$ on this simple definition, entirely free from any obscurity and from the admission of ideal numbers.†

Each system $\mathfrak{a}$ *of integers in a field* Ω, *possessing properties I and II, is called an* IDEAL OF THAT FIELD.

Divisibility of a number α by a number μ means that α is a number $\mu\omega$ in the principal ideal corresponding to the number μ and which can be conveniently denoted by $\mathfrak{o}(\mu)$ or $\mathfrak{o}\mu$. At the same time it follows from property II or theorem 2 that all the numbers in the principal ideal $\mathfrak{o}\alpha$ are also numbers in the principal ideal $\mathfrak{o}\mu$. Conversely, it is evident that α is certainly divisible by μ when all numbers in the ideal $\mathfrak{o}\alpha$, and hence α itself, are in the ideal $\mathfrak{o}\mu$. This leads us to establish the following notion of *divisibility*, not just for principal ideals, but for all ideals:

An ideal $\mathfrak{a}$ *is said to be* divisible *by an ideal* $\mathfrak{b}$, *or a multiple of* $\mathfrak{b}$, *and* $\mathfrak{b}$ *is said to be a divisor of* $\mathfrak{a}$, *when all numbers in the ideal* $\mathfrak{a}$ *are also in* $\mathfrak{b}$. *An ideal* $\mathfrak{p}$, *different from* $\mathfrak{o}$, *which has no divisors other than* $\mathfrak{o}$ *and* $\mathfrak{p}$, *is called a* prime ideal.‡

Divisibility of ideals, *which evidently includes that of numbers*, must at first be distinguished from the following notion of *multiplication* and the *product* of two ideals:

If α *runs through all the numbers in an ideal* $\mathfrak{a}$, *and* β *runs through all the numbers in an ideal* $\mathfrak{b}$, *then all the products of the form* $\alpha\beta$, *and all the sums of these products, form an ideal called the product of the ideals* $\mathfrak{a}$, $\mathfrak{b}$, *which we denote by* $\mathfrak{a}\mathfrak{b}$.§

One sees immediately, it is true, that the product $\mathfrak{a}\mathfrak{b}$ is divisible by $\mathfrak{a}$ and $\mathfrak{b}$, but establishing the complete connection between the notions of divisibility and multiplication of ideals succeeds only after we have vanquished the deep difficulties characteristic of the nature of the subject. This connection is essentially expressed by the following two theorems:

If the ideal $\mathfrak{c}$ *is divisible by the ideal* $\mathfrak{a}$, *then there is a unique ideal* $\mathfrak{b}$ *such that* $\mathfrak{a}\mathfrak{b} = \mathfrak{c}$.

---

† It is of course permissible, though not at all necessary, to let each *ideal* $\mathfrak{a}$ correspond to an *ideal number* which engenders it, if $\mathfrak{a}$ is not a principal ideal.

‡ Likewise the ideal number corresponding to the ideal $\mathfrak{a}\mathfrak{b}$ is said to be *divisible* by the ideal number corresponding to the ideal $\mathfrak{b}$, and corresponding to a prime ideal one has a prime ideal number.

§ The ideal number corresponding to the ideal $\mathfrak{a}\mathfrak{b}$ is called the *product* of the two ideal numbers corresponding to $\mathfrak{a}$ and $\mathfrak{b}$.

Each ideal different from o *is either a prime ideal, or uniquely expressible as a product of prime ideals.*

In the present memoir I confine myself to proving these results in a completely rigorous and synthetic way. This provides a proper *foundation* for the whole theory of ideals and decomposable forms, which offers to mathematicians an inexhaustible field of research. Of all the later developments, for which I refer to the exposition I have given in Dirichlet's *Vorlesungen über Zahlentheorie* and other memoirs still to appear, I have included here only the partition of ideals into *classes*, and the proof that the number of *classes of ideals* (or of classes of the corresponding forms) is finite. The first section contains only the propositions necessary for the present goal, extracted from an auxiliary theory, also important for other researches, which I shall expound fully elsewhere. The second section, which aims to clarify the general notions by numerical examples, can be omitted entirely. However, I have kept it because it may help in the understanding of the later sections, where the theory of integers in an arbitrary field of finite degree is developed from the above viewpoint. To do this one needs to borrow just the elements of the general theory of fields, a theory whose further development leads easily to the algebraic principles invented by Galois, which in their turn serve as a basis for deeper researches into the theory of ideals.

1

Auxiliary theorems from the theory of modules

As I have emphasised in the Introduction, we shall frequently have to consider systems of numbers closed under *addition* and *subtraction*. The general properties of such systems form a theory so extensive that it can also be used in other researches; nevertheless, for our purposes just the elements of this theory are sufficient. In order to avoid later interruption to the course of our exposition, and at the same time to make it easier to understand the scope of the concepts on which our theory of algebraic numbers is based, it seems appropriate to begin with a small number of very simple theorems, even though their interest lies mainly in their applications.

§1. Modules and their divisibility

1. A system $\mathfrak{a}$ of real or complex numbers will be called a *module* when all the sums and differences of these numbers also belong to $\mathfrak{a}$.

Thus if α is a particular number in the module $\mathfrak{a}$, all the numbers

$$\alpha + \alpha = 2\alpha, \quad 2\alpha + \alpha = 3\alpha, \quad \ldots,$$

$$\alpha - \alpha = 0, \quad 0 - \alpha = -\alpha, \quad -\alpha - \alpha = -2\alpha, \quad \ldots,$$

and consequently all numbers of the form $x\alpha$ also belong to $\mathfrak{a}$, where x runs through all the rational integers, that is, all the numbers

$$0, \pm 1, \pm 2, \pm 3, \ldots.$$

Such a system of numbers $x\alpha$ itself forms a module, which we denote by $[\alpha]$. Consequently, if a module includes a nonzero number then it includes an infinity of different numbers. It is also evident that the number zero, which is in each module, forms a module by itself.

2. A module $\mathfrak{a}$ will be called *divisible* by the module $\mathfrak{b}$ or a *multiple* of $\mathfrak{b}$, and $\mathfrak{b}$ a *divisor* of $\mathfrak{a}$, when all the numbers in the module $\mathfrak{a}$ are also in the module $\mathfrak{b}$.

The zero module is therefore a common multiple of all modules. Also, if α is a particular number in a module $\mathfrak{a}$, then the module $[\alpha]$ will be divisible by $\mathfrak{a}$. Moreover, it is evident that any module is divisible by itself, and that two modules $\mathfrak{a}$, $\mathfrak{b}$ divisible by each other are identical, which we shall denote by $\mathfrak{a} = \mathfrak{b}$. Finally, if each of the modules $\mathfrak{a}, \mathfrak{b}, \mathfrak{c}, \mathfrak{d}, \ldots$ is divisible by its immediate successor then it is clear that each will be divisible by all its successors.

3. Let $\mathfrak{a}$, $\mathfrak{b}$ be any two modules. The system $\mathfrak{m}$ of all the numbers that belong to both modules will itself be a module. It will be called the *least common multiple* of $\mathfrak{a}$, $\mathfrak{b}$ because each common multiple of $\mathfrak{a}$, $\mathfrak{b}$ is divisible by $\mathfrak{m}$.

Indeed, let μ, μ' be any two numbers in the system $\mathfrak{m}$, and hence in both $\mathfrak{a}$ and $\mathfrak{b}$. Each of the two numbers $\mu \pm \mu'$ will belong (by 1) not only to the module $\mathfrak{a}$ but also to the module $\mathfrak{b}$, and hence also to the system $\mathfrak{m}$, whence it follows that $\mathfrak{m}$ is a module. Since all members of this module $\mathfrak{m}$ are in $\mathfrak{a}$ and also in $\mathfrak{b}$, $\mathfrak{m}$ is a common multiple of $\mathfrak{a}$, $\mathfrak{b}$. Moreover, if the module $\mathfrak{m}'$ is any common multiple of $\mathfrak{a}$, $\mathfrak{b}$, and thus composed entirely of numbers belonging to both $\mathfrak{a}$ and $\mathfrak{b}$, then (by virtue of the definition of the system $\mathfrak{m}$) these numbers will also be in $\mathfrak{m}$, that is, $\mathfrak{m}'$ is divisible by $\mathfrak{m}$.

4. If α becomes equal in succession to all the numbers in a module $\mathfrak{a}$, and β to all the numbers in a module $\mathfrak{b}$, then the system $\mathfrak{d}$ of all numbers $\alpha + \beta$ will form a module. This module is called the *greatest common divisor* of $\mathfrak{a}$, $\mathfrak{b}$ because every common divisor of $\mathfrak{a}$, $\mathfrak{b}$ is also a divisor of $\mathfrak{d}$.

Indeed, any two numbers δ, δ' in the system $\mathfrak{d}$ can be put in the form $\delta = \alpha + \beta$, $\delta' = \alpha' + \beta'$ where α, α' belong to the module $\mathfrak{a}$ and β, β' to the module $\mathfrak{b}$, whence

$$\delta \pm \delta' = (\alpha \pm \alpha') + (\beta \pm \beta');$$

and, since the numbers $\alpha \pm \alpha'$ are in $\mathfrak{a}$ and the numbers $\beta \pm \beta'$ are in $\mathfrak{b}$, the numbers $\delta \pm \delta'$ also belong to the system $\mathfrak{d}$. That is, $\mathfrak{d}$ is a module. Since the number zero is in every module, all the numbers $\alpha = \alpha + 0$

of the module $\mathfrak{a}$ and all the numbers $\beta = 0 + \beta$ of the module $\mathfrak{b}$ belong to the module $\mathfrak{d}$. Consequently, the latter is a common divisor of $\mathfrak{a}$ and $\mathfrak{b}$. Also, if the module $\mathfrak{d}'$ is any common divisor of $\mathfrak{a}$, $\mathfrak{b}$, so that all the numbers in $\mathfrak{a}$ and all the numbers in $\mathfrak{b}$ are in $\mathfrak{d}'$ then (by virtue of 1) all the numbers $\alpha + \beta$, that is, all the numbers in the module $\mathfrak{d}$, also belong to the module $\mathfrak{d}'$. Thus $\mathfrak{d}$ is divisible by $\mathfrak{d}'$.

Having carried out these rigorous proofs, we need not explain further how the notions of least common multiple and greatest common divisor can be extended to any number (even an infinity) of modules. Nevertheless, it may be useful to justify the terminology chosen, by the following remark. If a, b are two particular rational integers, m their least common multiple and d their greatest common divisor, it follows from the elements of number theory that $[m]$ will be the least common multiple, and $[d]$ the greatest common divisor, of the modules $[a]$, $[b]$. In any case we shall soon see that the number-theoretic propositions relevant to this case can also be deduced from the theory of modules.

§2. Congruences and classes of numbers

1. Let $\mathfrak{a}$ be a module. Numbers ω, ω' will be called *congruent* or *incongruent* modulo $\mathfrak{a}$ according as their difference $\pm(\omega - \omega')$ is in $\mathfrak{a}$ or not. *Congruence* of the numbers ω, ω' with respect to the module $\mathfrak{a}$ will be indicated by the notation

$$\omega \equiv \omega' \pmod{\mathfrak{a}}.$$

We immediately deduce the following simple propositions, whose proofs we can omit:

If $\omega \equiv \omega' \pmod{\mathfrak{a}}$ and $\omega' \equiv \omega'' \pmod{\mathfrak{a}}$ then $\omega \equiv \omega'' \pmod{\mathfrak{a}}$.

If $\omega \equiv \omega' \pmod{\mathfrak{a}}$ and x is any rational integer then $x\omega \equiv x\omega' \pmod{\mathfrak{a}}$.

If $\omega \equiv \omega' \pmod{\mathfrak{a}}$ and $\omega'' \equiv \omega''' \pmod{\mathfrak{a}}$ then $\omega \pm \omega'' \equiv \omega' \pm \omega''' \pmod{\mathfrak{a}}$.

If $\omega \equiv \omega' \pmod{\mathfrak{a}}$ and $\mathfrak{d}$ is a divisor of $\mathfrak{a}$, then $\omega \equiv \omega' \pmod{\mathfrak{d}}$.

If $\omega \equiv \omega' \pmod{\mathfrak{a}}$ and $\omega \equiv \omega' \pmod{\mathfrak{b}}$ then $\omega \equiv \omega' \pmod{\mathfrak{m}}$, where $\mathfrak{m}$ is the least common multiple of $\mathfrak{a}$, $\mathfrak{b}$.

2. The first of the preceding theorems leads to the notion of a class of numbers relative to a module $\mathfrak{a}$, by which we mean the set of those numbers congruent to a particular number, and hence to each other, modulo $\mathfrak{a}$. Such a class modulo $\mathfrak{a}$ is completely determined by giving a

single member, and each member can be regarded as a *representative* of the whole class. The numbers in the module $\mathfrak{a}$, for example, form such a class, represented by the number zero.

Now if $\mathfrak{b}$ is a second module we can always choose in $\mathfrak{b}$ a finite or infinite number of numbers

$$(\beta_r) \qquad\qquad \beta_1, \quad \beta_2, \quad \beta_3, \quad \ldots,$$

in such a way that each number in $\mathfrak{b}$ is congruent modulo $\mathfrak{a}$ to one of these numbers, and to only one. Such a system of numbers β_r in the module $\mathfrak{b}$, which are mutually incongruent modulo $\mathfrak{a}$, but which represent all classes having members in $\mathfrak{b}$, I call a *complete system of representatives of the module* $\mathfrak{b}$ *modulo the module* $\mathfrak{a}$, and the number of numbers β_r, or of the classes they represent, will be denoted by $(\mathfrak{b}, \mathfrak{a})$, when it is finite. If, on the contrary, the number of representatives β_r is infinite, it will be convenient to assign the value zero to the symbol $(\mathfrak{b}, \mathfrak{a})$. A deeper examination of such a system (β_r) of representatives now leads to the following theorem:

3. Let $\mathfrak{a}$, $\mathfrak{b}$ be any two modules, with least common multiple $\mathfrak{m}$ and greatest common divisor $\mathfrak{d}$. Any complete system of representatives of the module $\mathfrak{b}$ modulo $\mathfrak{a}$ will at the same time be a complete system of representatives of the module $\mathfrak{b}$ modulo $\mathfrak{m}$, and for the module $\mathfrak{d}$ modulo $\mathfrak{a}$; consequently we have

$$(\mathfrak{b}, \mathfrak{a}) = (\mathfrak{b}, \mathfrak{m}) = (\mathfrak{d}, \mathfrak{a}).$$

First of all, it is evident that any numbers β, β' in the module $\mathfrak{b}$ which are congruent modulo $\mathfrak{a}$ are congruent modulo $\mathfrak{m}$, because $\beta - \beta'$ is in $\mathfrak{a}$ as well as in $\mathfrak{b}$, and hence also in $\mathfrak{m}$. Now, since each number β in the module $\mathfrak{b}$ is congruent to one of the representatives β_r modulo $\mathfrak{a}$, and hence also modulo $\mathfrak{m}$, and since any two different representatives are incongruent modulo $\mathfrak{a}$ and hence also modulo $\mathfrak{m}$; these numbers β_r in $\mathfrak{b}$ form a complete system of representatives of the module $\mathfrak{b}$ modulo $\mathfrak{m}$. The second part is proved in absolutely the same way: since $\mathfrak{b}$ is divisible by $\mathfrak{d}$, the numbers β_r are likewise in $\mathfrak{d}$ and, by hypothesis, they are incongruent modulo $\mathfrak{a}$. And, since each number δ in $\mathfrak{d}$ is of the form $\alpha + \beta$, where α is in $\mathfrak{a}$ and β is in $\mathfrak{b}$, we have

$$\delta = \alpha + \beta \equiv \beta \pmod{\mathfrak{a}},$$

and, since β and consequently δ is congruent to one of the numbers β_r modulo $\mathfrak{a}$, the numbers β_r form a complete system of representatives for the module $\mathfrak{d}$ modulo $\mathfrak{a}$. Q.E.D.

If $\mathfrak{b}$ is divisible by $\mathfrak{a}$ then $(\mathfrak{b}, \mathfrak{a}) = 1$, because all numbers in $\mathfrak{b}$ are $\equiv 0$ (mod $\mathfrak{a}$). Conversely, if $(\mathfrak{b}, \mathfrak{a}) = 1$ then $\mathfrak{b}$ will be divisible by $\mathfrak{a}$, since all numbers in $\mathfrak{b}$ are congruent to each other and hence $\equiv 0$ (mod $\mathfrak{a}$). We evidently have at the same time $\mathfrak{m} = \mathfrak{b}$, $\mathfrak{d} = \mathfrak{a}$.

4. If $\mathfrak{d}$ is a divisor of $\mathfrak{a}$ and at the same time a multiple of $\mathfrak{c}$, moreover if β_r runs through the representatives of $\mathfrak{b}$ modulo $\mathfrak{a}$ and if γ_s runs through the representatives of $\mathfrak{c}$ modulo $\mathfrak{b}$, then the numbers $\beta_r + \gamma_s$ form a complete system of representatives of the module $\mathfrak{c}$ modulo $\mathfrak{a}$, and consequently

$$(\mathfrak{c}, \mathfrak{a}) = (\mathfrak{c}, \mathfrak{b})(\mathfrak{b}, \mathfrak{a}).$$

In the first place, all the numbers $\beta_r + \gamma_s$ belong to the module $\mathfrak{c}$, since β_r is in $\mathfrak{b}$ and hence also in $\mathfrak{c}$, and γ_s is likewise in $\mathfrak{c}$.

In the second place, they are all incongruent modulo $\mathfrak{a}$. In fact if we let β', β'' be particular values of β_r, and let γ', γ'' be particular values of γ_s, then the hypothesis $\beta' + \gamma' \equiv \beta'' + \gamma''$ (mod $\mathfrak{a}$) implies, since $\mathfrak{a}$ is divisible by $\mathfrak{b}$ and $\beta' \equiv \beta'' \equiv 0$ (mod $\mathfrak{b}$), that $\gamma' \equiv \gamma''$ (mod $\mathfrak{b}$). However, since γ', γ'' are particular members of the series γ_s, any two of which are incongruent modulo $\mathfrak{b}$, we must have $\gamma' = \gamma''$ and hence the hypothesis above becomes $\beta' \equiv \beta''$ (mod $\mathfrak{a}$). Now since β', β'' likewise are particular members of the series β_r, any two of which are incongruent modulo $\mathfrak{a}$, we must have $\beta' = \beta''$, which proves the assertion above.

In the third place, it remains to see that each number γ in $\mathfrak{c}$ is congruent to one of the numbers $\beta_r + \gamma_s$ modulo $\mathfrak{a}$. In fact, since each number γ is congruent to one of the numbers γ_s modulo $\mathfrak{b}$, we can choose $\gamma = \beta + \gamma_s$, where β is a number in the module $\mathfrak{b}$. Also, since each of these numbers β is congruent to one of the numbers β_r modulo $\mathfrak{a}$, we can choose $\beta = \alpha + \beta_r$, where α is a number in the module $\mathfrak{a}$. We then have

$$\gamma = \beta + \gamma_s = \alpha + \beta_r + \gamma_s \equiv \beta_r + \gamma_s \ (\text{mod } \mathfrak{a}).$$

Q.E.D.

5. Let $\mathfrak{m}$ be the least common multiple, and $\mathfrak{d}$ the greatest common divisor, of two modules $\mathfrak{a}$, $\mathfrak{b}$, and let ρ, σ be given numbers. The system of two congruences

$$\omega \equiv \rho \ (\text{mod } \mathfrak{a}), \qquad \omega \equiv \sigma \ (\text{mod } \mathfrak{b})$$

has a common root if and only if

$$\rho \equiv \sigma \ (\text{mod } \mathfrak{d}),$$

and all such numbers ω form a class of numbers modulo the module $\mathfrak{m}$.

If there is a number ω satisfying the two congruences then the numbers $\omega - \rho$, $\omega - \sigma$ will be in $\mathfrak{a}$, $\mathfrak{b}$ respectively, and since the latter are both in $\mathfrak{d}$, the difference $\rho - \sigma$ of the numbers will likewise be in $\mathfrak{d}$. That is, the above condition $\rho \equiv \sigma \ (\text{mod } \mathfrak{d})$ is necessary. Conversely, if this condition is satisfied then (by virtue of the definition of $\mathfrak{d}$ in §1,4) there is a number α in $\mathfrak{a}$ and a number β in $\mathfrak{b}$ whose sum $\alpha + \beta = \rho - \sigma$, hence the number $\omega = \rho - \alpha = \sigma + \beta$ satisfies the two congruences. Thus the condition is also sufficient. Moreover, if ω' is a number satisfying the same conditions as ω, then $\omega' - \omega$ will also be in both $\mathfrak{a}$ and $\mathfrak{b}$, and hence also in $\mathfrak{m}$, which means that $\omega' \equiv \omega \ (\text{mod } \mathfrak{m})$. Conversely, each number ω' in the class represented by ω modulo $\mathfrak{m}$ will satisfy the congruences. Q.E.D.

§3. Finitely generated modules

1. Let $\beta_1, \beta_2, \beta_3, \ldots, \beta_n$ be particular numbers. All the numbers

$$\beta = y_1\beta_1 + y_2\beta_2 + y_3\beta_3 + \cdots + y_n\beta_n,$$

where $y_1, y_2, y_3, \ldots, y_n$ are arbitrary rational integers, evidently form a module, which we call a *finitely generated*† module $[\beta_1, \beta_2, \beta_3, \ldots, \beta_n]$. The complex of constants $\beta_1, \beta_2, \beta_3, \ldots, \beta_n$ will be called a *basis*‡ of the module.

This module $[\beta_1, \beta_2, \ldots, \beta_n]$ is evidently the greatest common divisor of the n finitely generated modules $[\beta_1], [\beta_2], \ldots, [\beta_n]$. It is easy to see that each multiple of a finitely generated module is itself a finitely generated module, but here I confine myself to proving the following fundamental theorem, which will later have important applications.

2. If all the numbers in a finitely generated module $\mathfrak{b} = [\beta_1, \beta_2, \ldots, \beta_n]$ can be transformed into the members of a module $\mathfrak{a}$ by multiplication by nonzero rational numbers, then the least common multiple $\mathfrak{m}$ of $\mathfrak{a}$

---

† Dedekind calls them "finite". (Translator's note.)
‡ Note that Dedekind's basis elements need not be independent. However, he requires them to be independent for fields (§15). (Translator's note.)

and $\mathfrak{b}$ will be a finitely generated module, and one can choose a system of $\frac{1}{2}(n+1)n$ rational integers a such that the n numbers

$$\mu_1 = a_1'\beta_1,$$
$$\mu_2 = a_1''\beta_1 + a_2''\beta_2,$$

$$\cdots\cdots\cdots\cdots\cdots\cdots$$

$$\mu_n = a_1^{(n)}\beta_1 + a_2^{(n)}\beta_2 + a_3^{(n)}\beta_3 + \cdots + a_n^{(n)}\beta_n,$$

form a basis of $\mathfrak{m}$, and at the same time

$$(\mathfrak{b}, \mathfrak{a}) = (\mathfrak{b}, \mathfrak{m}) = a_1' a_2'' a_3''' \cdots a_n^{(n)}.$$

By hypothesis there are n nonzero fractions

$$\frac{s_1}{t_1}, \quad \frac{s_2}{t_2}, \quad \frac{s_3}{t_3}, \quad \ldots, \quad \frac{s_n}{t_n},$$

whose numerators and denominators are rational integers, such that the n products

$$\frac{s_1}{t_1}\beta_1, \quad \frac{s_2}{t_2}\beta_2, \quad \frac{s_3}{t_3}\beta_3, \quad \ldots, \quad \frac{s_n}{t_n}\beta_n$$

belong to the module $\mathfrak{a}$. Since members of a module $\mathfrak{a}$ are changed into other members of $\mathfrak{a}$ when multiplied by rational integers $t_1, t_2, t_3, \ldots, t_n$ (§1,1), the products $s_1\beta_1, s_2\beta_2, s_3\beta_3, \ldots, s_n\beta_n$ likewise belong to $\mathfrak{a}$, and if s denotes the absolute value of the product $s_1 s_2 s_3 \cdots s_n$, the numbers $s\beta_1, s\beta_2, s\beta_3, \ldots, s\beta_n$, and consequently all products $s\beta$, belong to the module $\mathfrak{a}$, where β denotes any number in the module $\mathfrak{b}$.

Now let ν be a particular index from the sequence $1, 2, \ldots, n$. Among the numbers in the module $[\beta_1, \beta_2, \ldots, \beta_\nu]$ divisible by $\mathfrak{b}$ let

$$\mu_\nu' = y_1\beta_1 + y_2\beta_2 + \cdots + y_\nu\beta_\nu$$

denote those belonging to module $\mathfrak{a}$ and hence also to module $\mathfrak{b}$, for example $s\beta_\nu$. Among these numbers μ_ν' there will be at least one number

$$\mu_\nu = a_1^{(\nu)}\beta_1 + a_2^{(\nu)}\beta_2 + \cdots + a_\nu^{(\nu)}\beta_\nu,$$

for which y takes the *smallest positive* value $a_\nu^{(\nu)}$. One can then see that, in *all* the numbers μ_ν', the coefficient y is divisible by $a_\nu^{(\nu)}$. This is because one can always put

$$y_\nu = x_\nu a_\nu^{(\nu)} + y_\nu'$$

where x_ν and y_ν' are rational integers and the latter satisfies the condition†

$$0 \le y_\nu' < a_\nu^{(\nu)}.$$

† Which is the foundation of the theory of division for rational integers.

Then if we put

$$y_1' = y_1 - x_\nu a_1^{(\nu)}, \quad y_2' = y_2 - x_\nu a_2^{(\nu)}, \quad \ldots, \quad y_{\nu-1}' = y_{\nu-1} - x_\nu a_{\nu-1}^{(\nu)}$$

the number

$$\mu_\nu' - x_\nu \mu_\nu = y_1' \beta_1 + y_2' \beta_2 + \cdots + y_{\nu-1}' \beta_{\nu-1} + y_\nu' \beta_\nu$$

belongs to both the module $[\beta_1, \beta_2, \ldots, \beta_\nu]$ and the module $\mathfrak{m}$, since μ_ν' and μ_ν are in $\mathfrak{m}$. But since (by definition of the μ_ν) the coefficient of β_ν is less than $a_\nu^{(\nu)}$ and at the same time positive, it is necessary that $y_\nu' = 0$, and hence that $y_\nu = x_\nu a_\nu^{(\nu)}$ be divisible by $a_\nu^{(\nu)}$, as required. At the same time

$$\mu_\nu' - x_\nu \mu_\nu = \mu_{\nu-1}'$$

becomes a number in $[\beta_1, \beta_2, \ldots, \beta_{\nu-1}]$ and also in $\mathfrak{m}$, or else becomes zero in the case $\nu = 1$.

It follows easily that the n numbers μ_ν, obtained by putting $\nu = n, n-1, \ldots, 2, 1$ in succession, enjoy the properties enunciated in the theorem. Each number μ in the module $\mathfrak{m}$, that is, each number μ_n' in both $\mathfrak{a}$ and $\mathfrak{b} = [\beta_1, \beta_2, \ldots, \beta_n]$, is of the form

$$\mu = \mu_{n-1}' + x_n \mu_n$$

where x_n is a rational integer and μ_{n-1}' is a number belonging to the two modules $\mathfrak{a}$ and $[\beta_1, \beta_2, \ldots, \beta_{n-1}]$, and hence also to the module $\mathfrak{m}$. Each number μ_{n-1}' of this nature is of the form

$$\mu_{n-1}' = \mu_{n-2}' + x_{n-1} \mu_{n-1},$$

where x_{n-1} is a rational integer and μ_{n-2}' is a number belonging to the two modules $\mathfrak{a}$ and $[\beta_1, \beta_2, \ldots, \beta_{n-1}]$, and so on. Finally, each number μ_1' belonging to the two modules $\mathfrak{a}$ and $[\beta_1]$ is of the form

$$\mu_1' = x_1 \mu_1$$

where x_1 is a rational integer. Thus it is proved that each number μ of the module $\mathfrak{m}$ can be represented in the form

$$\mu = x_1 \mu_1 + x_2 \mu_2 + \cdots + x_n \mu_n,$$

where $x_1, x_2, \ldots, x_n$ are rational integers. Conversely, since an arbitrarily chosen system of rational integers $x_1, x_2, \ldots, x_n$ certainly produces a number μ in the module $\mathfrak{m}$, because $\mu_1, \mu_2, \ldots, \mu_n$ themselves belong to $\mathfrak{m}$, the n numbers $\mu_1, \mu_2, \ldots, \mu_n$ form a basis of the module $\mathfrak{m}$.

To prove the last part of the theorem we have to consider all the numbers

$$\beta' = z'_1\beta_1 + z'_2\beta_2 + \cdots + z'_n\beta_n$$

of the module $\mathfrak{b}$ for which the rational integers $z'_1, z'_2, \ldots, z'_n$ satisfy the n conditions

$$0 \leq z'_\nu < a_\nu^{(\nu)}.$$

We shall show that these numbers β', the number of which is evidently equal to $a'_1 a''_2 \cdots a_\nu^{(\nu)}$, form a complete system of representatives of the module $\mathfrak{b}$ modulo $\mathfrak{m}$ (or $\mathfrak{a}$).

In the first place, all the numbers β' in the module $\mathfrak{b}$ are incongruent modulo $\mathfrak{m}$. If

$$z'_1\beta_1 + \cdots + z'_n\beta_n \equiv z''_1\beta_1 + \cdots + z''_n\beta_n \pmod{\mathfrak{a}},$$

then the numbers $z''_1, z''_2, \ldots, z''_n$ satisfy the same n conditions as the numbers $z'_1, z'_2, \ldots, z'_n$. Then if the n differences

$$z'_n - z''_n, \quad z'_{n-1} - z''_{n-1}, \quad \ldots, \quad z'_2 - z''_2, \quad z'_1 - z''_1$$

are not all zero, let $z'_\nu - z''_\nu$ be the first of them with a nonzero value, a value which we can assume to be positive by symmetry, and which is also $< a_\nu^{(\nu)}$ since both numbers z'_ν and z''_ν are $< a_\nu^{(\nu)}$. Then the difference

$$(z'_1 - z''_1)\beta_1 + \cdots + (z'_\nu - z''_\nu)\beta_\nu$$

is evidently a number μ'_ν in $\mathfrak{a}$ and $[\beta_1, \beta_2, \ldots, \beta_n]$ for which the coefficient of β_ν is positive and $< a_\nu^{(\nu)}$, contrary to the definition of the number μ_ν. Thus any two different systems of n numbers $z'_1, z'_2, \ldots, z'_n$, which satisfy the conditions above, also produce two numbers β' in the module $\mathfrak{b}$ which are incongruent modulo $\mathfrak{a}$.

In the second place, it is easy to see that an arbitrary number

$$\beta = z_1\beta_1 + z_2\beta_2 + \cdots + z_n\beta_n$$

in the module $\mathfrak{b}$ is congruent modulo $\mathfrak{a}$ (or $\mathfrak{m}$) to one of the numbers β' since, if $z_1, z_2, \ldots, z_n$ are given, it is clear that we can successively choose n rational integers

$$x_n, x_{n-1}, \ldots, x_2, x_1$$

so that the n numbers

$$z'_n = z_n + a_n^{(n)} x_n,$$
$$z'_{n-1} = z_{n-1} + a_{n-1}^{(n)} x_n + a_{n-1}^{(n-1)} x_{n-1},$$

$$\cdots\cdots\cdots\cdots\cdots\cdots\cdots\cdots\cdots\cdots\cdots$$

$$z'_2 = z_2 + a_2^{(n)} x_n + a_2^{(n-1)} x_{n-1} + \cdots + a''_2 x_2,$$

$$z'_1 = z_1 + a_1^{(n)} x_n + a_1^{(n-1)} x_{n-1} + \cdots + a''_1 x_2 + a'_1 x_1,$$

satisfy the n conditions $0 \le z'_\nu < a_\nu^{(\nu)}$. If we now put

$$\beta' = z'_1 \beta_1 + z'_2 \beta_2 + \cdots + z'_n \beta_n,$$

we have

$$\beta' = \beta + x_1 \mu_1 + x_2 \mu_2 + \cdots + x_n \mu_n,$$

and hence $\beta \equiv \beta'$ (mod $\mathfrak{m}$).   Q.E.D.

§4. *Irreducible systems*

1. A system of n numbers $\alpha_1, \alpha_2, \ldots, \alpha_n$ will be called an *irreducible system*, and its members will be called *independent*, when the sum

$$\alpha = x_1 \alpha_1 + x_2 \alpha_2 + \cdots + x_n \alpha_n$$

is nonzero for any system of rational numbers $x_1, x_2, \ldots, x_n$ which are not all zero. It then follows that any two different systems of rational numbers $x_1, x_2, \ldots, x_n$ produce unequal sums α. In the contrary case, that is, when there is a system of rational numbers $x_1, x_2, \ldots, x_n$, not all zero, for which the sum α is zero, then the system of numbers $\alpha_1, \alpha_2, \ldots, \alpha_n$ will be called *reducible*, and the numbers themselves will be called *dependent* on each other. If one wants to retain this terminology in the case $n = 1$, a single number evidently forms a reducible or irreducible system according as it is zero or not. The definition above easily yields the following theorems, whose number can be increased enormously, on the *determinants* of rational numbers.

2. If the n numbers $\alpha_1, \alpha_2, \ldots, \alpha_n$ are independent, then the n numbers

$$\alpha'_1 = c'_1 \alpha_1 + c'_2 \alpha_2 + \cdots + c'_n \alpha_n,$$

$$\alpha'_2 = c''_1 \alpha_1 + c''_2 \alpha_2 + \cdots + c''_n \alpha_n,$$

$$\cdots\cdots\cdots\cdots\cdots\cdots\cdots\cdots\cdots\cdots\cdots$$

$$\alpha'_n = c_1^{(n)} \alpha_1 + c_2^{(n)} \alpha_2 + \cdots + c_n^{(n)} \alpha_n,$$

whose n^2 coefficients c are rational numbers, form an irreducible or reducible system according as the determinant

$$C = \sum \pm c_1' c_2'' \cdots c_n^{(n)}$$

is nonzero or not.

Since $\alpha_1, \alpha_2, \ldots, \alpha_n$ are independent, the sum

$$x_1 \alpha_1' + x_2 \alpha_2' + \cdots + x_n \alpha_n' = \alpha',$$

where $x_1, x_2, \ldots, x_n$ are arbitrary rational numbers, not all zero, cannot vanish unless we simultaneously have

$$c_1' x_1 + c_1'' x_2 + \cdots + c_1^{(n)} x_n = 0,$$
$$c_2' x_2 + c_2'' x_2 + \cdots + c_2^{(n)} x_n = 0,$$
$$\cdots\cdots\cdots\cdots\cdots\cdots\cdots\cdots\cdots$$
$$c_n' x_1 + c_n'' x_2 + \cdots + c_n^{(n)} x_n = 0,$$

which is impossible when C is nonzero. Hence the numbers $\alpha_1', \alpha_2', \ldots, \alpha_n'$ are independent in that case. But if we have $C = 0$ there is always a system of rational numbers $x_1, x_2, \ldots, x_n$ satisfying the preceding equations, and not all zero. This is seen immediately when all the n^2 coefficients c vanish. If this is not the case then, among the minor determinants of C that do not vanish there will be one, say

$$\sum \pm c_1' c_2'' \cdots c_r^{(r)},$$

of *maximal* rank $r < n$ such that the minor determinants of higher degree vanish. In this case, as we know, the last $n - r$ of the equations above will be consequences of the preceding r, and we can put these r equations in the form

$$x_1 = p_{r+1}' x_{r+1} + \cdots + p_n' x_n,$$
$$\cdots\cdots\cdots\cdots\cdots\cdots\cdots\cdots\cdots$$
$$x_r = p_{r+1}^{(r)} x_{r+1} + \cdots + p_n^{(r)} x_n,$$

where the $r(n-r)$ coefficients p are rational numbers. Now if we give the $n - r$ quantities $x_{r+1}, \ldots, x_n$ arbitrary rational values then not only can we ensure that they are not all zero, the quantities $x_1, \ldots, x_r$ will likewise take rational values. Thus we have a system of n rational numbers $x_1, x_2, \ldots, x_n$, not all zero, for which the sum α' is zero. Hence in this case the n numbers $\alpha_1', \alpha_2', \ldots, \alpha_n'$ are dependent.   Q.E.D.

3. If the n independent numbers $\alpha_1, \alpha_2, \ldots, \alpha_n$ form one basis of a module $\mathfrak{a}$, and $\alpha_1', \alpha_2', \ldots, \alpha_n'$ form another, then we have

$$\alpha_\nu' = c_1^{(\nu)}\alpha_1 + c_2^{(\nu)}\alpha_2 + \cdots + c_n^{(\nu)}\alpha_n,$$

where the n^2 coefficients c are rational integers whose determinant is ± 1, and consequently the numbers $\alpha_1', \alpha_2', \ldots, \alpha_n'$ are also independent.

In fact, since the numbers α_ν' are in the module $\mathfrak{a} = [\alpha_1, \alpha_2, \ldots, \alpha_n]$ there are in any case n equations of the preceding form, in which the coefficients c are rational integers. Conversely, since the n numbers α_ν are in the module $\mathfrak{a} = [\alpha_1', \alpha_2', \ldots, \alpha_n']$ there are also n equations of the form

$$\alpha_\nu = e_1^{(\nu)}\alpha_1' + e_2^{(\nu)}\alpha_2' + \cdots + e_n^{(\nu)}\alpha_n',$$

whose coefficients e are likewise rational integers. Substituting in them the first n equations for the n numbers α_ν', and bearing in mind that the n numbers α_ν form an irreducible system, it follows that the sum

$$e_1^{(\nu)}c_{\nu'}' + e_2^{(\nu)}c_{\nu'}'' + \cdots + e_n^{(\nu)}c_{\nu'}^{(n)} = 1 \text{ or } 0$$

according as the indices ν, ν' are equal or not. Then the product of the determinants

$$\sum \pm c_1' c_2'' \cdots c_n^{(n)} \cdot \sum \pm e_1' e_2'' \cdots e_n^{(n)} = 1,$$

and, since each factor is a rational integer,

$$\sum \pm c_1' c_2'' \cdots c_n^{(n)} = \sum \pm e_1' e_2'' \cdots e_n^{(n)} = \pm 1.$$

Q.E.D.

Conversely, it is clear that $[\alpha_1', \alpha_2', \ldots, \alpha_n'] = [\alpha_1, \alpha_2, \ldots, \alpha_n]$ when there are n equations of the form

$$\alpha_\nu' = c_1^{(\nu)}\alpha_1 + \cdots + c_n^{(\nu)}\alpha_n,$$

where the coefficients c are rational integers whose determinant $= \pm 1$.

4. If the n independent numbers $\beta_1, \ldots, \beta_n$ form the basis of a module $\mathfrak{b}$, and if n numbers $\alpha_1, \ldots, \alpha_n$, forming the basis of a module $\mathfrak{a}$, depend on them via n equations of the form

$$\alpha_\nu = b_1^{(\nu)}\beta_1 + \ldots + b_n^{(\nu)}\beta_n,$$

where the coefficients b are rational integers whose determinant B is nonzero, then the number of classes is

$$(\mathfrak{b}, \mathfrak{a}) = \pm B.$$

Now each of the numbers $\beta_1, \ldots, \beta_n$, and hence each number β, of the module $[\beta_1, \beta_2, \ldots, \beta_n]$ can be changed into a member of the module $\mathfrak{a}$ by multiplying by a nonzero rational number B. It follows, since $\mathfrak{a}$ is divisible by $\mathfrak{b}$ and hence also equal to the least common multiple of $\mathfrak{a}$ and $\mathfrak{b}$, that $\mathfrak{a}$ has (by §4,3) a basis of n numbers of the form

$$\alpha'_\nu = a_1^{(\nu)}\beta_1 + a_2^{(\nu)}\beta_2 + \ldots + a_\nu^{(\nu)}\beta_\nu,$$

whose coefficients a are rational integers chosen so that

$$(\mathfrak{b}, \mathfrak{a}) = a'_1 a''_2 \cdots a_n^{(n)} = \sum \pm a'_1 a''_2 \cdots a_n^{(n)}.$$

Moreover, since the n numbers $\alpha_1, \ldots, \alpha_n$ likewise form a basis of the module $\mathfrak{a}$ and since (by 2) each of these two systems of n numbers is irreducible, because we assume this of the system $\beta_1, \ldots, \beta_n$, we then have (by 3) n equations of the form

$$\alpha'_\nu = c_1^{(\nu)}\alpha_1 + \cdots + c_n^{(\nu)}\alpha_n,$$

with rational integer coefficients c whose determinant

$$\sum \pm c'_1 c''_2 \cdots c_n^{(n)} = \pm 1.$$

By replacing the numbers $\alpha_1, \ldots, \alpha_n$ by their expressions above in terms of the n independent numbers $\beta_1, \ldots, \beta_n$ we see, by comparison with the preceding expressions for the numbers α'_ν in terms of the same numbers $\beta_1, \ldots, \beta_n$, that

$$\alpha_{\nu'}^{(\nu)} = c_1^{(\nu)}b'_{\nu'} + c_2^{(\nu)}b''_{\nu'} + \cdots + c_n^{(\nu)}b_{\nu'}^{(n)},$$

and consequently

$$\sum \pm a'_1 \cdots a_n^{(n)} = \sum \pm c'_1 \cdots c_n^{(n)} \cdot \sum \pm b'_1 \cdots b_n^{(n)}.$$

Thus we have $(\mathfrak{b}, \mathfrak{a}) = \pm B$.   Q.E.D.

This important theorem can easily be extended (and even more simply by means of the theorem below) to the more general case where the coefficients b are *fractional* rational numbers. One then obtains the theorem

$$(\mathfrak{b}, \mathfrak{a}) = \pm B(\mathfrak{a}, \mathfrak{b}),$$

and each of the two numbers of classes, $(\mathfrak{a}, \mathfrak{b})$ and $(\mathfrak{b}, \mathfrak{a})$, can be determined by a simple rule involving the determinant B and all its minor determinants.

5. If only n among the m numbers $\alpha_1, \alpha_2, \ldots, \alpha_m$ forming a basis

of the module $\mathfrak{a}$ are independent, then $\mathfrak{a}$ has a basis consisting of n independent numbers $\alpha_1', \alpha_2', \ldots, \alpha_n'$.

The hypothesis of this theorem will evidently be satisfied whenever all the m numbers $\alpha_1, \ldots, \alpha_m$ are expressible in terms of n independent numbers $\omega_1, \ldots, \omega_n$, as

$$\alpha_\mu = r_1^{(\mu)} \omega_1 + r_2^{(\mu)} \omega_2 + \cdots + r_n^{(\mu)} \omega_n,$$

where the system of coefficients

$$(r) \quad \begin{cases} r_1', & r_2', & \ldots, & r_n', \\ r_1'', & r_2'', & \ldots, & r_n'', \\ \ldots & \ldots & \ldots, & \ldots, \\ r_1^{(m)}, & r_2^{(m)}, & \ldots, & r_n^{(m)} \end{cases}$$

consists of rational numbers, at least one of whose

$$\frac{m(m-1)\ldots(m-n+1)}{1 \cdot 2 \cdots \cdot n}$$

$n \times n$ partial determinants R is nonzero. Otherwise, any n of the m numbers a_μ would be dependent. Conversely, it follows from the hypothesis of the theorem, that the m numbers α_μ can always be expressed in terms of n independent numbers ω_ν, by choosing the latter to be, for example, n numbers among the m numbers α_μ which form an irreducible system. Then, since the $n+1$ numbers $\alpha_\mu, \omega_1, \ldots, \omega_n$ are dependent there is an equation, for each index μ, of the form

$$x_0 \alpha_\mu + x_1 \omega_1 + x_2 \omega_2 + \cdots + x_n \omega_n = 0,$$

with rational coefficients x not all zero. Moreover, since $\omega_1, \omega_2, \ldots, \omega_n$ are independent, x_0 must be nonzero, and consequently α_μ can be represented, in the manner indicated, in terms of the numbers ω_ν. Finally, since the m numbers α_μ include the n numbers ω_ν, at least one of the determinants R will be nonzero.

I shall therefore assume that the m numbers α_μ are represented, in the manner indicated, in terms of the n independent numbers ω_ν, and I shall show that, no matter how the numbers ω_ν are chosen, there is a basis of the module $\mathfrak{a} = [\alpha_1, \alpha_2, \ldots, \alpha_m]$ consisting of n numbers α_ν' of the form

$$\alpha_\nu' = c_1^{(\nu)} \omega_1 + c_2^{(\nu)} \omega_2 + \cdots + c_\nu^{(\nu)} \omega_\nu,$$

with rational coefficients c. To do this I remark first that we can evidently choose a positive integer k so that the mn products $k r_\nu^{(\mu)}$ are integers. If we now put

$$\omega_1 = k\beta_1, \quad \omega_2 = k\beta_2, \quad \ldots, \quad \omega_n = k\beta_n,$$

and express the numbers α_μ in terms of the numbers β_n it follows that the module $\mathfrak{a} = [\alpha_1, \alpha_2, \cdots, \alpha_m]$ is divisible by the module $\mathfrak{b} = [\beta_1, \beta_2, \ldots, \beta_n]$, and hence it is the least common multiple of $\mathfrak{a}$, $\mathfrak{b}$. Moreover, since the n numbers β_ν become the n numbers ω_ν when multiplied by k and the latter, when multiplied by a nonzero determinant R, become numbers of the form

$$x_1\alpha_1 + x_2\alpha_2 + \cdots + x_m\alpha_m,$$

with rational coefficients x, it is clear that each number β in the module $\mathfrak{b}$, when multiplied by a nonzero rational, itself becomes a number in the module $\mathfrak{a}$. It follows from this (by §3,2) that the least common multiple of the two modules $\mathfrak{a}$, $\mathfrak{b}$ has a basis consisting of n numbers of the form

$$\alpha'_\nu = \alpha_1^{(\nu)}\beta_1 + \alpha_2^{(\nu)}\beta_2 + \cdots + \alpha_\nu^{(\nu)}\beta_\nu,$$

with rational coefficients a for which the product $a'_1 a''_2 \cdots a_n^{(n)}$ is nonzero. If we now re-express the numbers β_ν in terms of the n numbers ω_ν we can conclude that the assertion above is true, which at the same time proves the theorem.

6. To the preceding proof I add the following remarks. Since the m numbers α_μ form a basis for the module $\mathfrak{a}$ just as much as the n numbers α'_ν, there are m equations of the form

$$\alpha_\mu = p_1^{(\mu)}\alpha'_1 + p_2^{(\mu)}\alpha'_2 + \cdots + p_n^{(\mu)}\alpha'_n,$$

and n equations of the form

$$\alpha'_\nu = q'_\nu\alpha_1 + q''_\nu\alpha_2 + \cdots + q_\nu^{(m)}\alpha_m,$$

where the $2mn$ coefficients p and q are all rational *integers*. By substituting the first expressions in the second, and bearing in mind that the n numbers α'_ν are independent, we deduce that the sum

$$q'_\nu p'_{\nu'} + q''_\nu p''_{\nu'} + \cdots + q_\nu^{(m)} p_{\nu'}^{(m)} = 1 \text{ or } 0,$$

according as the m indices ν, ν' from the series $1, 2, \ldots, n$ are equal or not. Then if P denotes any $n \times n$ partial determinant formed from the system of coefficients (p), and if Q denotes any determinant formed similarly from the system of coefficients (q), then we know that the sum

$$\sum PQ,$$

taken over all combinations of n different upper indices, is equal to 1, and consequently the determinants P have no common divisor. Conversely, this property of the determinants P is necessary if the n numbers α'_ν, and also the m numbers

$$\alpha_\mu = p_1^{(\mu)}\alpha'_1 + \cdots + p_n^{(\mu)}\alpha'_n,$$

are to form a basis of the module $\mathfrak{a}$.

A system of coefficients such as (p) is evidently just a special case of the preceding coefficients (r). Now since the n numbers α'_ν can likewise be represented in the form

$$\alpha'_\nu = e_1^{(\nu)}\omega_1 + e_2^{(\nu)}\omega_2 + \cdots + e_n^{(\nu)}\omega_n,$$

with n^2 rational coefficients e whose determinant

$$E = \sum \pm e'_1 e''_2 \cdots e_n^{(n)}$$

is nonzero, we get

$$r_\nu^{(\mu)} = p_1^{(\mu)}e'_\nu + p_2^{(\mu)}e''_\nu + \cdots + p_n^{(\mu)}e_\nu^{(n)}.$$

Consequently, the two determinants R, P corresponding to systems of coefficients (r), (p) satisfy the relation

$$R = PE.$$

The problem of finding all the systems (p) corresponding to a given system (r) can be solved in the most comprehensive and elegant manner by generalising a method applied by Gauss† in the special case in which one utilises identities between the partial determinants. However, this would lead us too far away from our present position, and I am content to have shown the *existence* of a system such as (p). One sees immediately (from 3) that one can derive from it all other systems (p) by composition with all possible systems of n^2 rational integers with determinant ± 1.

In practice, that is, in any case where the coefficients r are given numerically and, without loss of generality, as *integers*, we arrive most promptly at the goal by a chain of elementary transformations. These are based on the evident proposition that a module $[\alpha_1, \alpha_2, \ldots, \alpha_m]$ is not altered when we replace α_1, for example, by the number $\alpha_1 + x\alpha_2$, where x is any rational integer. The partial determinants R^0 corresponding to all combinations of n numbers from the new basis

$$\alpha_1^0 = \alpha_1 + x\alpha_2, \quad \alpha_2^0 = \alpha_2, \quad \alpha_3^0 = \alpha_3, \quad \ldots, \quad \alpha_m^0 = \alpha_m,$$

† *Disquisitiones Arithmeticae*, art. 234, 236, 279.

and to the new system of coefficients (r^0), coincide in part with the determinants R corresponding to the old basis

$$\alpha_1 = \alpha_1^0 - x\alpha_1^0, \quad \alpha_2 = \alpha_2^0, \quad \alpha_3 = \alpha_3^0, \quad \ldots, \quad \alpha_m = \alpha_m^0.$$

They are of the form $R_1^0 = R_1 + xR_2$, from which we easily deduce that the greatest common divisor E of the determinants R is the same as that of the determinants R^0. Thus the determinants R^0 cannot simultaneously vanish. We shall now use these basis transformations of the module $\mathfrak{a}$ as follows:

The m coefficients $r_n^{(\mu)}$ of the number ω_n cannot all be zero, since then all the determinants R would be zero. Now if *two* of these coefficients, say r_n' and r_n'', are nonzero, and if $|r_n'| \geq |r_n''|$, then we can choose a rational integer x such that $|r_n' + xr_n''| < |r_n''|$.† The elementary transformation above therefore gives us a new basis in which all the m coefficients $r_n^{(\mu)}$, except the first, r_n', remain the same, and this single coefficient is replaced by one of *smaller* absolute value. By repetition of this procedure we necessarily arrive at a basis in which all but one of the m coefficients of ω_n are zero. We denote the member of the basis for which the latter coefficient $a_n^{(n)}$ is nonzero by

$$\alpha_n' = a_1^{(n)}\omega_1 + a_2^{(n)}\omega_2 + \cdots + a_n^{(n)}\omega_n,$$

and we keep it fixed in all subsequent transformations of the basis. The partial determinants corresponding to the actual basis either vanish, or are of the form $Sa_n^{(n)}$, where S is an $(n-1) \times (n-1)$ partial determinant corresponding to an arbitrary combination of $n-1$ of the $m-1$ members of the basis other than α_n', and which is formed from the $(n-1)^2$ coefficients corresponding to $\omega_1, \omega_2, \ldots, \omega_{n-1}$. Since the determinants S cannot all be zero, we now proceed with these $m-1$ members of the basis, treating ω_{n-1} as we previously did ω_n in working with the m numbers a_μ of the original basis. If we continue these transformations then we finally obtain a basis of $\mathfrak{a}$ consisting of n numbers $\alpha_1', \alpha_2', \ldots, \alpha_{n-1}', \alpha_n'$ of the form

$$\alpha_\nu' = a_1^{(\nu)}\omega_1 + a_2^{(\nu)}\omega_2 + \cdots + a_n^{(\nu)}\omega_\nu,$$

and $m-n = s$ numbers $\alpha_1'', \alpha_2'', \ldots, \alpha_s''$ which are all zero, and which can therefore be omitted. The n nonzero coefficients $a_\nu^{(\nu)}$ can be taken to be positive, since α_ν' can be replaced by $-\alpha_\nu'$ without alteration of the

---

† Here again is the same principle which is fundamental in the theory of rational integers.

module, and their product $a_1' a_2'' \cdots a_n^{(n)}$ is evidently the greatest common divisor E of all the partial determinants R.

In this way we obtain a second proof of the important theorem (5), and at the same time it is evident that, by composition of the successive transformations and their inversion, we can find the system of coefficients (p) as well as a system of coefficients (q). In fact, one first obtains m equations of the form

$$\alpha_\mu = \sum_\nu p_\nu^{(\mu)} \alpha_\nu' + \sum_\sigma h_\sigma^{(\mu)} \alpha_\sigma''$$

or, since the s numbers α_σ'' are zero,

$$\alpha_\mu = \sum_\nu p_\nu^{(\mu)} \alpha_\nu';$$

and since the determinant of each of the substitutions or transformations is equal to 1, the $m \times m$ determinant

$$\begin{vmatrix} p_1' & \cdots & p_n' & h_1' & \cdots & h_s' \\ \cdot\cdot & \cdots & \cdot\cdot & \cdot\cdot & \cdots & \cdot\cdot \\ p_1^{(m)} & \cdots & p_n^{(m)} & h_1^{(m)} & \cdots & h_s^{(m)} \end{vmatrix} = \sum PH = 1,$$

since the quantities H are $s \times s$ determinants complementary to the determinants P, and formed from the system of coefficients (h). By inversion, one obtains the adjoint determinant

$$\begin{vmatrix} q_1' & \cdots & q_n' & k_1' & \cdots & k_s' \\ \cdot\cdot & \cdots & \cdot\cdot & \cdot\cdot & \cdots & \cdot\cdot \\ q_1^{(m)} & \cdots & q_n^{(m)} & k_1^{(m)} & \cdots & k_s^{(m)} \end{vmatrix} = \sum QK = 1,$$

where K denotes the determinant complementary to Q, and if P, Q are corresponding determinants then we know that $H = Q$, $K = P$. At the same time we obtain n equations of the form

$$\alpha_\nu' = \sum_\mu q_\nu^{(\mu)} \alpha_\mu$$

and s equations of the form

$$\alpha_\sigma'' = \sum_\mu k_\sigma^{(\mu)} \alpha_\mu = 0.$$

The latter equations are a new expression of the original supposition that only n of the m numbers α_μ are independent, and we have been able to base all this study on such a system of s equations.

One can generally shorten the calculation itself by carrying out several elementary transformations simultaneously. Suppose for example that $m = 4$, $n = 2$, whence $s = 2$, and

$$\alpha_1 = 21\omega_1, \quad \alpha_2 = 7\omega_1 + 7\omega_2, \quad \alpha_3 = 9\omega_1 - 3\omega_2, \quad \alpha_4 = 8\omega_1 + 2\omega_2,$$

giving

$$(r) \qquad \begin{cases} r'_1 = 21, & r''_1 = 7, & r'''_1 = 9, & r''''_1 = 8, \\ r'_2 = 0, & r''_2 = 7, & r'''_2 = -3, & r''''_2 = 2. \end{cases}$$

We obtain the six partial determinants

$$(R) \qquad \begin{cases} R_{1,2} = 147, & R_{1,3} = -63, & R_{1,4} = 42, \\ R_{3,4} = 42, & R_{2,4} = -42, & R_{2,3} = -84, \end{cases}$$

using the abbreviation

$$r_1^{(\mu)} r_2^{(\mu')} - r_1^{(\mu')} r_2^{(\mu)} = R_{\mu,\mu'}.$$

These determinants satisfy the identity

$$R_{1,2} R_{3,4} - R_{1,3} R_{2,4} + R_{1,4} R_{2,3} = 0.$$

Now since the smallest nonzero coefficient of ω_2 is in α_4 we form the new basis

$$\beta_1 = \alpha_1 = 21\omega_1, \qquad\qquad \beta_2 = \alpha_2 - 3\alpha_4 = -17\omega_1 + \omega_2,$$
$$\beta_3 = \alpha_3 + 2\alpha_4 = 25\omega_1 + \omega_2, \quad \beta_4 = \alpha_4 = 8\omega_1 + 2\omega_2,$$

whence, conversely

$$\alpha_1 = \beta_1, \quad \alpha_2 = \beta_2 + 3\beta_4, \quad \alpha_3 = \beta_3 - 2\beta_1, \quad \alpha_4 = \beta_4.$$

Now since ω_2 has 1 as its smallest nonzero coefficient, in β_2 for example, we form the third basis

$$\gamma_1 = \beta_1 = 21\omega_1, \qquad\qquad \gamma_2 = \beta_2 = -17\omega_1 + \omega_2,$$
$$\gamma_3 = -\beta_2 + \beta_3 = 42\omega_1, \quad \gamma_4 = -2\beta_2 + \beta_4 = 42\omega_1,$$

whence, conversely,

$$\beta_1 = \gamma_1, \quad \beta_2 = \gamma_2, \quad \beta_3 = \gamma_2 + \gamma_3, \quad \beta_4 = 2\gamma_2 + \gamma_4.$$

At this stage, since γ_2 is the only number in which ω_2 has a nonzero coefficient, and since γ_1 is the number, amongst the other three, in which ω_1 has its least coefficient 21, we form the fourth basis

$$\delta_1 = \gamma_1 = 21\omega_1, \qquad\qquad \delta_2 = \gamma_2 = -17\omega_1 + \omega_2,$$
$$\delta_3 = -2\gamma_1 + \gamma_3 = 0, \quad \delta_4 = -2\gamma_1 + \gamma_4 = 0,$$

whence, conversely,

$$\gamma_1 = \delta_1, \quad \gamma_2 = \delta_2, \quad \gamma_3 = 2\delta_1 + \delta_3, \quad \gamma_4 = 2\delta_1 + \delta_4 = 0.$$

Since $\delta_3 = \delta_4 = 0$ the transformation is completed, and successive substitutions give

$$
\begin{array}{rclcl}
\alpha_1 & = & \delta_1 & = & \delta_1, \\
\alpha_2 & = & 6\delta_1 \; +7\delta_2 \qquad\qquad +3\delta_4 & = & 6\delta_1 \; +7\delta_2, \\
\alpha_3 & = & -2\delta_1 \; -3\delta_2 \; +\delta_3 \; -2\delta_4 & = & -2\delta_1 \; -3\delta_2, \\
\alpha_4 & = & 2\delta_1 \; +2\delta_2 \qquad\qquad +\delta_4 & = & 2\delta_1 \; +2\delta_2,
\end{array}
$$

and conversely

$$
\begin{array}{rclcl}
\delta_1 & = & \alpha_1 & = & 21\omega_1, \\
\delta_2 & = & \alpha_2 \qquad\qquad -3\alpha_4 & = & -17\omega_1 \; +\omega_2, \\
\delta_3 & = & -2\alpha_1 \; -\alpha_2 \; +\alpha_3 \; +5\alpha_4 & = & 0, \\
\delta_4 & = & -2\alpha_1 \; -2\alpha_2 \qquad +7\alpha_4 & = & 0.
\end{array}
$$

Since $\delta_1, \delta_2, \delta_3, \delta_4$ are the quantities which, in the general theory, we have denoted by $\alpha_1', \alpha_2', \alpha_1'', \alpha_2''$, we have

$$(p) \qquad \begin{cases} p_1' = 1, \quad p_1'' = 6, \quad p_1''' = -2, \quad p_1'''' = 2, \\ p_2' = 0, \quad p_2'' = 7, \quad p_2''' = -3, \quad p_2'''' = 2. \end{cases}$$

Thus we obtain, for the determinants proportional to the R,

$$(P) \qquad \begin{cases} P_{1,2} = 7, \quad P_{1,3} = -3, \quad P_{1,4} = 2, \\ P_{3,4} = 2, \quad P_{2,4} = -2, \quad P_{2,3} = -4, \end{cases}$$

and likewise

$$(q) \qquad \begin{cases} q_1' = 1, \quad q_1'' = 0, \quad q_1''' = 0, \quad q_1'''' = 0, \\ q_2' = 0, \quad q_2'' = 1, \quad q_2''' = 0, \quad q_2'''' = -3, \end{cases}$$

and

$$(Q) \qquad \begin{cases} Q_{1,2} = 1, \quad Q_{1,3} = 0, \quad Q_{1,4} = -3, \\ Q_{3,4} = 0, \quad Q_{2,4} = 0, \quad Q_{2,3} = 0. \end{cases}$$

Finally, from the systems of coefficients

$$(h) \qquad \begin{cases} h_1' = 0, \quad h_1'' = 0, \quad h_1''' = 1, \quad h_1'''' = 0, \\ h_2' = 0, \quad h_2'' = 3, \quad h_2''' = -2, \quad h_2'''' = 1, \end{cases}$$

and

$$(k) \qquad \begin{cases} k_1' = -2, \quad k_1'' = -1, \quad k_1''' = 1, \quad k_1'''' = 5, \\ k_2' = -2, \quad k_2'' = -2, \quad k_2''' = 0, \quad k_2'''' = 7, \end{cases}$$

we derive the determinants $H_{\mu,\mu'} = Q_{\mu,\mu'}$ and $K_{\mu,\mu'} = P_{\mu,\mu'}$ complementary to $P_{\mu,\mu'}$ and $Q_{\mu,\mu'}$ respectively,

$$(H) \begin{cases} H_{1,2} = h_1''' h_2'''' - h_1'''' h_2''', & H_{1,3} = h_1'''' h_2'' - h_1'' h_2'''', & H_{1,4} = h_1'' h_2''' - h_1''' h_2'', \\ H_{3,4} = h_1' h_2'' - h_1'' h_2', & H_{2,4} = h_1''' h_2' - h_1' h_2''', & H_{2,3} = h_1' h_2'''' - h_1'''' h_2', \end{cases}$$

$$(K) \begin{cases} K_{1,2} = k_1''' k_2'''' - k_1'''' k_2''', & K_{1,3} = k_1'''' k_2'' - k_1'' k_2'''', & K_{1,4} = k_1'' k_2''' - k_1''' k_2'', \\ K_{3,4} = k_1' k_2'' - k_1'' k_2', & K_{2,4} = k_1''' k_2' - k_1' k_2''', & K_{2,3} = k_1' k_2'''' - k_1'''' k_2', \end{cases}$$

and the treatment of the example is complete.

To conclude, I remark that applying this to the case $n = 1$ leads to the fundamental theorem on the greatest common divisor of an arbitrary number of rational integers, the foundation for the whole divisibility theory of these numbers.

The researches in this first chapter have been expounded in a special form suited to our goal, but it is clear that they do not cease to be true when the Greek letters denote not only numbers, but any objects of study, any two of which α, β produce a determinate third element $\gamma = \alpha + \beta$ of the same type, under a commutative and uniformly invertible operation (composition), taking the place of addition. The module $\mathfrak{a}$ becomes a *group* of elements, the composites of which all belong to the same group. The rational integer coefficients indicate how many times an element contributes to the generation of another.

2

Germ of the theory of ideals

In this chapter I propose, as indicated in the Introduction, to explain in a particular case the nature of the phenomenon that led Kummer to the creation of *ideal numbers*, and I shall use the same example to explain the concept of *ideal* introduced by myself, and that of the multiplication of ideals.

§5. *The rational integers*

The theory of numbers is at first concerned exclusively with the system of rational integers $0, \pm 1, \pm 2, \pm 3, \ldots$, and it will be worthwhile to recall in a few words the important laws that govern this domain. Above all, it should be recalled that these numbers are closed under addition, subtraction and multiplication, that is, the sum, difference and product of any two members in this domain also belong to the domain. The theory of *divisibility* considers the combination of numbers under multiplication. The number a is said to be divisible by the number b when $a = bc$, where c is also a rational integer. The number 0 is divisible by any number; the two units ± 1 divide all numbers, and they are the only numbers that enjoy this property. If a is divisible by b, then $\pm a$ will also be divisible by $\pm b$, and consequently we can restrict ourselves to the consideration of positive numbers. Each positive number, different from unity, is either a *prime* number, that is, a number divisible only by itself and unity, or else a *composite* number. In the latter case we can always express it as a product of prime numbers and – which is the most important thing – in only one way. That is, the system of prime numbers occurring as factors in this product is completely determined by giving the number of times a designated prime number occurs as a factor. This property depends essentially on the theorem that a prime divides a product of two factors only when it divides one of the factors.

The simplest way to prove these fundamental propositions of number theory is based on the algorithm taught by Euclid, which serves to find the greatest common divisor of two numbers.† This procedure, as we know, is based on repeated application of the theorem that, for a positive number m, any number z can be expressed in the form $qm + r$, where q and r are also integers and r is *less* than m. It is for this reason that the procedure always halts after a finite number of divisions.

The notion of *congruence* of numbers was introduced by Gauss.‡ Two numbers z, z' are called congruent modulo the modulus m, written

$$z \equiv z' \pmod{m},$$

when the difference $z - z'$ is divisible by m. In the contrary case z and z' are called *incongruent* modulo m. If we arrange the numbers in classes, with two numbers in the same class§ only if they are congruent modulo m, then we easily conclude from the theorem recalled above that the number of classes is finite and equal to the absolute value of the modulus m. This also follows from the studies of the preceding chapter, since the definition of congruence in Chapter 1 contains that of Gauss as a special case. The system $\mathfrak{o}$ of all rational integers is identical with the finitely generated module $[1]$, and likewise the system $\mathfrak{m}$ of all numbers divisible by m is identical with $[m]$. The congruence of two numbers modulo m coincides with congruence modulo the system $\mathfrak{m}$. Thus (by §3,2 or §4,4) the number of classes is $(\mathfrak{o}, \mathfrak{m}) = \pm m$.

§6. The complex integers of Gauss

The first and greatest step in the generalisation of these notions was made by Gauss, in his second memoir on biquadratic residues, when he transported them to the domain of complex integers $x + yi$, where x and y are any rational integers and i is $\sqrt{-1}$, that is, a root of the irreducible quadratic equation $i^2 + 1 = 0$. The numbers in this domain are closed under addition, subtraction and multiplication, and consequently we can define divisibility for these numbers in the same way as for rational

---

† See, for example, the *Vorlesungen über Zahlentheorie* of Dirichlet.
‡ *Disquisitiones Arithmeticae*, art. 1.
§ The word *class* seems to have been employed by Gauss first à propos of *complex* numbers. (*Theoria residuorum biquadraticorum*, II, art. 42.)

numbers. One can establish very simply, as Dirichlet showed in a very elegant manner,¶ that the general propositions on the composition of numbers from primes continue to hold in this new domain, as a result of the following remark. If we define the *norm* $N(w)$ of a number $w = u+vi$, where u and v are any rational numbers, to be the product $u^2 + v^2$ of the two conjugate numbers $u + vi$ and $u - vi$, then the norm of a product will be equal to the product of the norms of the factors, and it is also clear that for any given w we can choose a complex *integer* q such that $N(w - q) \leq 1/2$. If now we let z and m be any complex integers, with m nonzero, it follows by taking $w = z/m$ that we can put $z = qm + r$ where q and r are complex integers such that $N(r) < N(m)$. We can then find a greatest common divisor of any two complex integers by a finite number of divisions, exactly as for rational numbers, and the proofs of the general laws of divisibility for rational integers can be applied word for word in the domain of complex integers. There are four units, ± 1, $\pm i$, that is, four numbers which divide all numbers, and whose norm is consequently 1. Every other nonzero number is either a composite number, so called when it is the product of two factors, neither of which is a unit, or else it is a prime, and such a number cannot divide a product unless it divides at least one of the factors. Every composite number can be expressed uniquely as a product of prime numbers, provided of course the four associated primes $\pm q$, $\pm qi$ are regarded as representatives of the same prime number q. The set of all prime numbers q in the domain of complex integers consists of:

1. All the rational prime numbers (taken positively) of the form $4n+3$;

2. The number $1 + i$, dividing the rational prime $2 = (1 + i)(1 - i) = -i(1 + i)^2$;

3. The factors $a + bi$ and $a - bi$ of each rational prime p of the form $4n + 1$ with norm $a^2 + b^2 = p$.

The existence of the primes $a \pm bi$ just mentioned, which follows immediately from the celebrated theorem of Fermat on the equation $p = a^2 + b^2$, and which likewise implies that theorem, can now be derived without the help of the theorem, with marvellous ease. It is a splendid example of the extraordinary power of the principles we have reached through generalisation of the notion of integer.

Congruence of complex integers modulo a given number m of the same kind can also be defined in absolutely the same way as in the theory of rational numbers. Numbers z, z' are called congruent modulo m, written

¶ *Recherches sur les formes quadratiques à coefficients et à indéterminées complexes* (Crelle's Journal, 24).

$z \equiv z'$ (mod m), when $z - z'$ is divisible by m. If we arrange the numbers into classes, with numbers being in the same class or not according as they are congruent or incongruent modulo m, then the total number of different classes will be finite and equal to $N(m)$. This follows very easily from the researches of the first chapter, since the system $\mathfrak{o}$ of all complex integers $x + yi$ forms a finitely generated module $[1, i]$ and likewise the system $\mathfrak{m}$ of all the numbers $m(x + yi)$ divisible by m forms the module $[m, mi]$, whose basis is related to that of $\mathfrak{o}$ by two equations of the form

$$m = a \cdot 1 + b \cdot i, \quad mi = -b \cdot 1 + a \cdot i.$$

Consequently we have (§4,4)

$$(\mathfrak{o}, \mathfrak{m}) = \begin{vmatrix} a & b \\ -b & a \end{vmatrix} = N(m).$$

§7. The domain $\mathfrak{o}$ of numbers $x + y\sqrt{-5}$

There are still other numerical domains which can be treated in absolutely the same manner. For example, let θ be any root of any of the five equations

$$\theta^2 + \theta + 1 = 0, \quad \theta^2 + \theta + 2 = 0,$$

$$\theta^2 + 2 = 0, \quad \theta^2 - 2 = 0, \quad \theta^2 - 3 = 0,$$

and let x, y be any rational integers. Then the numbers $x + y\theta$ form a corresponding numerical domain. In each of these domains it is easy to see that one can find the greatest common divisor of two numbers by a finite number of divisions, so that one immediately has general laws of divisibility agreeing with those for rational numbers, even though there happen to be an infinite number of units in the last two examples.

On the other hand, this method is not applicable to the domain $\mathfrak{o}$ of integers

$$\omega = x + y\theta$$

where θ is a root of the equation

$$\theta^2 + 5 = 0,$$

and x, y again take all rational integer values. Here we encounter the phenomenon which suggested to Kummer the creation of ideal numbers, and which we shall now describe in detail by means of examples.

The numbers ω of the domain o we shall now be concerned with are closed under addition, subtraction and multiplication, and we therefore define the notions of divisibility and congruence of numbers exactly as before. Also, if we define the norm $N(\omega)$ of a number $\omega = x + y\theta$ to be the product $x^2 + 5y^2$ of the two conjugate numbers $x \pm y\theta$, then the norm of a product will be equal to the product of the norms of the factors. And if μ is a particular nonzero number we conclude, just as before, that $N(\mu)$ expresses how many mutually incongruent numbers there are modulo μ. If μ is a unit, and hence divides all numbers, then we must have $N(\mu) = 1$ and therefore $\mu = \pm 1$.

A number (different from zero and ± 1) is called *decomposable* when it is the product of two factors, neither of which is a unit. In the contrary case the number is called *indecomposable*. Then it follows from the theorem on the norm that each decomposable number can be expressed as the product of a finite number of indecomposable factors. However, in infinitely many cases an entirely new phenomenon presents itself here, namely, the same number is susceptible to several, essentially different, representations of this kind. The simplest examples are the following. It is easy to convince oneself that each of the following fifteen numbers is indecomposable.

$$
\begin{array}{llll}
a = 2, & b = 3, & c = 7; & \\
b_1 = -2 + \theta, & b_2 = -2 - \theta; & c_1 = 2 + 3\theta, & c_2 = 2 - 3\theta; \\
d_1 = 1 + \theta, & d_2 = 1 - \theta; & e_1 = 3 + \theta, & e_2 = 3 - \theta; \\
f_1 = -1 + 2\theta, & f_2 = -1 - 2\theta; & g_1 = 4 + \theta, & g_2 = 4 - \theta.
\end{array}
$$

In fact, for a rational prime p to be decomposable, and hence of the form $\omega\omega'$, it is necessary that $N(p) = p^2 = N(\omega)N(\omega')$, and since ω, ω' are not units we must have $p = N(\omega) = N(\omega')$, that is, p must be representable by the binary quadratic form $x^2 + 5y^2$. But the three prime numbers 2, 3, 7 cannot be represented in this way, as one sees from the theory of these forms,† or else by a small number of direct trials. They are therefore indecomposable. It is easy to show the same thing, similarly, for the other twelve numbers, whose norms are products of two of these three primes. However, despite the indecomposability of these fifteen numbers, there are numerous relations between their products, which can all be deduced from the following:

(1) $$ab = d_1 d_2, \qquad b^2 = b_1 b_2, \qquad ab_1 = d_1^2,$$

---

† See Dirichlet's *Vorlesungen über Zahlentheorie*, §71.

(2) $\qquad\qquad ac = e_1 e_2, \qquad c^2 = c_1 c_2, \qquad ac_1 = e_1^2,$

(3) $\qquad bc = f_1 f_2 = g_1 g_2, \qquad af_1 = d_1 e_1, \qquad ag_1 = d_1 e_2.$

In each of these ten relations, the same number is represented in two or three *different* ways as a product of indecomposable numbers. Thus one sees that an indecomposable number may very well divide a product without dividing any of its factors. Such an indecomposable number therefore does not possess the property which, in the theory of rational numbers, is characteristic of a *prime number.*

If we imagine for a moment that the fifteen preceding numbers are *rational* integers then, by the general laws of divisibility, we easily deduce from the relations (1) that there are decompositions of the form†

$$a = \mu\alpha^2, \qquad d_1 = \mu\alpha\beta_1, \qquad d_2 = \mu\alpha\beta_2,$$
$$b = \mu\beta_1\beta_2, \qquad b_1 = \mu\beta_1^2, \qquad b_2 = \mu\beta_2^2,$$

and from the relations (2) that there are decompositions of the form

$$a = \mu'\alpha_2', \qquad e_1 = \mu'\alpha'\gamma_1, \qquad e_2 = \mu'\alpha'\gamma_2,$$
$$c = \mu'\gamma_1\gamma_2, \qquad c_1 = \mu'\gamma_1^2, \qquad c_2 = \mu'\gamma_2^2,$$

where all the Greek letters denote rational integers. And it follows immediately, by virtue of the equation $\mu\alpha^2 = \mu'\alpha'^2$, that the four numbers f_1, f_2, g_1, g_2 appearing in the relations (3) will likewise be *integers.* These decompositions are simplified if we make the additional assumption that a is prime to b and c, since this implies $\mu = \mu' = 1$, $\alpha = \alpha'$ and hence the fifteen numbers can be expressed as follows, in terms of five numbers α, β_1, β_2, γ_1, γ_2:

---

† Since these decompositions do not seem obvious to me, I include the following
proof of the consequences of (1) as an example. Note first that $ab_1 = d_1^2$ and
$b_1 b_2 = b^2$ are both squares. Suppose that

$$a = \mu\alpha^2, \qquad b_1 = \mu_1\beta_1^2, \qquad b_2 = \mu_2\beta_2^2,$$

where μ, μ_1, μ_2 are squarefree. Then $ab_1 = \mu\mu_1\alpha^2\beta_1^2$ is not a square unless
$\mu = \mu_1$. Similarly, $b_1 b_2$ is not a square unless $\mu_1 = \mu_2$. Thus in fact $\mu = \mu_1 = \mu_2$
and hence

$$a = \mu\alpha^2, \qquad b_1 = \mu\beta_1^2, \qquad b_2 = \mu\beta_2^2.$$

Forming products of these, we get

$$d_1^2 = ab_1 = \mu^2\alpha^2\beta_1^2 \Rightarrow d_1 = \mu\alpha\beta_1,$$
$$d_2^2 = ab_2 = \mu^2\alpha^2\beta_2^2 \Rightarrow d_2 = \mu\alpha\beta_2,$$
$$b^2 = b_1 b_2 = \mu^2\beta_1^2\beta_2^2 \Rightarrow b = \mu\beta_1\beta_2,$$

which completes the proof of the decompositions claimed by Dedekind. (Translator's note.)

(4) $\begin{cases} a = \alpha^2, & b = \beta_1\beta_2, & c = \gamma_1\gamma_2; \\ b_1 = \beta_1^2, & b_2 = \beta_2^2; & c_1 = \gamma_1^2, & c_2 = \gamma_2^2; \\ d_1 = \alpha\beta_1, & d_2 = \alpha\beta_2; & e_1 = \alpha\gamma_1, & e_2 = \alpha\gamma_2; \\ f_1 = \beta_1\gamma_1, & f_2 = \beta_2\gamma_2; & g_1 = \beta_1\gamma_2, & g_2 = \beta_2\gamma_1. \end{cases}$

Now even though our fifteen numbers are in reality indecomposable, the remarkable thing is that they behave, in all questions of divisibility in the domain o, exactly as if they were composed, in the manner indicated above, of five different *prime numbers* α, β_1, β_2, γ_1, γ_2. In a moment I shall explain in detail what these relations between numbers mean.

§8. Role of the number 2 in the domain o

Let me begin by remarking that, in the theory of rational integers, one can recognise the essential constitution of a number *without effecting its decomposition* into prime factors, observing only how it behaves as a *divisor*. If we know, for example, that a positive number a does not divide a product of two squares unless at least one of the squares is divisible by a, then we can conclude with certainty that a is either 1, a prime or the square of a prime. It is likewise certain that a number a must contain at least one square factor, other than unity, when we can prove the existence of a number not divisible by a, whose square is divisible by a. Thus if we can ascertain that both these two properties hold for a, then we can conclude with certainty that a is the *square of a prime number*.

We shall now examine the behaviour, in this sense, of the number 2 in our domain o of numbers $\omega = x + y\theta$. Since any two conjugate numbers are congruent modulo 2 we have

$$\omega^2 \equiv N(\omega) \pmod 2,$$

and hence also $\omega^2\omega'^2 \equiv N(\omega)N(\omega') \pmod 2$. Now, if the number 2 is to divide the product $\omega^2\omega'^2$, and hence also the product of the two *rational* numbers $N(\omega)$, $N(\omega')$, it is necessary for at least one of these norms, and hence also for at least one of the two squares ω^2, ω'^2, to be divisible by 2. Moreover, if we take x, y to be any two odd numbers, then we obtain a number $\omega = x + y\theta$ not divisible by 2, and whose square is divisible by 2. Having regard to the preceding remarks on rational numbers, we then say that the number 2 behaves in our domain o as though it were the square of a prime number α.

Although such a prime number α does not actually exist in the domain $\mathfrak{o}$, it is by no means necessary to introduce it, since in fact Kummer managed in similar circumstances with great success by taking such a number α to be an *ideal number*, and we are allowing ourselves to be guided by analogy with the theory of rational numbers to define the presence of the number α in terms of *existing* numbers ω of the domain $\mathfrak{o}$. Now, when a rational number a is known to be the square of a prime α we can easily judge, *without bringing in* α, whether α is a factor of an arbitrary rational integer z, and how many times. It is clear that z is divisible by α^n if and only if z^2 is divisible by a^n. Thus we extend the criterion to the case we are interested in by saying that a number ω of the domain $\mathfrak{o}$ is *divisible* by the nth *power* α^n of the ideal prime number α when ω^2 is divisible by 2^n. Experience will show that this definition is very luckily† chosen, because it leads to a mode of expression in perfect harmony with the laws of the theory of rational numbers.

It follows first, for $n = 1$, that a number $\omega = x + y\theta$ is divisible by α if and only if $N(\omega)$ is an even number, and consequently

(α) $x \equiv y \pmod 2$.

The number ω is *not* divisible by α when $N(\omega)$ is an odd number, and consequently $x \equiv 1+y \pmod 2$. From this we get the theorem expressing the character of the ideal number α as a prime number:

"The product of numbers not divisible by α is also not divisible by α."

As far as higher powers of α are concerned, we first conclude from the definition that a number ω divisible by α^n is also divisible by all lower powers of α, because a number ω^2 divisible by 2^n is also divisible by all lower powers of 2. We now have to find, when ω is nonzero, the *highest* power α^m of α that divides ω, that is, the highest power of 2 that divides ω^2. Let s be the exponent of the highest power of 2 that divides ω itself. We have

$$\omega = 2^s \omega_1 = 2^s(x_1 + y_1\theta),$$

and at least one of the two rational integers x_1, y_1 will be odd. If both are odd, ω_1 will be divisible by α and we shall have

$$\omega_1^2 = x_1^2 - 5y_1^2 + 2x_1 y_1 \theta = 2\omega_2,$$

† Luckily, since, for example, trying analogously to determine the role of the number 2 in the domain of numbers $x + y\sqrt{-3}$ leads to complete failure. Later we shall clearly see the reason for this phenomenon.

where $\omega_2 = x_2 + y_2\theta$ is not divisible by α, because x_2 is even and y_2 is odd. But if one of the two numbers x_1, y_1 is even, so that the other is odd, then ω_1 and consequently ω_1^2 will not be divisible by α. Thus in the first case $m = 2s + 1$, in the second case $m = 2s$, but in both cases $\omega^2 = 2^m\omega'$ where ω' is a number not divisible by α. We see at the same time that m is also the exponent of the highest power of 2 that divides the norm $N(\omega)$. We therefore have the theorem:

"The exponent of the highest power of α that divides a product is equal to the sum of the exponents of the highest powers of α that divide the factors".

It is likewise evident that each number ω divisible by α^{2n} is also divisible by 2^n since, if the exponent denoted above by s is $< n$, then the numbers $2s$, $2s + 1$ and hence also m will be $< 2n$, contrary to hypothesis. It follows immediately from the definition that, conversely, each number divisible by 2^n is also divisible by α^{2n}.

Since the number $1 + \theta$ is divisible by α, but not by α^2, we easily see, with the help of the preceding theorem, that the congruence $\omega^2 \equiv 0$ (mod 2^n) that serves to define divisibility of the number ω by α^n can be replaced by the congruence

$$(\alpha^n) \qquad\qquad \omega(1 + \theta)^n \equiv 0 \;(\text{mod } 2^n),$$

which has the advantage of containing the number ω only to the *first* power.

§9. *Role of the numbers* 3 *and* 7 *in the domain* o

When all the quantities appearing in the equations (4) of §7 are *rational* integers, and if at the same time a is prime to b and c, then it is evident that a rational integer z will be divisible by β_1, β_2, γ_1, γ_2 according as it satisfies the corresponding congruences

$$zd_2 \equiv 0, \quad zd_1 \equiv 0 \quad (\text{mod } b),$$
$$ze_2 \equiv 0, \quad ze_1 \equiv 0 \quad (\text{mod } c).$$

These congruences have the peculiarity that they do not involve the numbers β_1, β_2, γ_1, γ_2 themselves, and it is for precisely this reason that they are appropriate for introducing the four ideal numbers β_1, β_2, γ_1, γ_2 in the context of numbers in the domain o. We say that a number

$\omega = x + y\theta$ is *divisible* by one of these four numbers if ω is a root of the corresponding congruence

$$(1 - \theta)\omega \equiv 0, \quad (1 + \theta)\omega \equiv 0 \quad (\text{mod } 3),$$
$$(3 - \theta)\omega \equiv 0, \quad (3 + \theta)\omega \equiv 0 \quad (\text{mod } 7).$$

Multiplication converts these congruences to the following:

(β_1)	$x \equiv y \quad (\text{mod } 3),$
(β_2)	$x \equiv -y \quad (\text{mod } 3),$
(γ_1)	$x \equiv 3y \quad (\text{mod } 7),$
(γ_2)	$x \equiv -3y \quad (\text{mod } 7),$

concerning which we add the following remarks.

Each of these conditions can be satisfied by one of the numbers $\omega = 1 + \theta$, $1 - \theta$, $3 + \theta$, $3 - \theta$, and the number in question does not satisfy any of the other three, so that it is legitimate to say that the four ideal numbers are *all different*. Moreover, since every number ω divisible by β_1 and β_2 is also divisible by 3, since $x \equiv y \equiv -y \equiv 0 \pmod{3}$ in that case, and since conversely every number divisible by 3 is also divisible by β_1 and β_2, we ought to regard 3 as the least common multiple of the ideal numbers β_1, β_2, by analogy with the theory of rational numbers. But each of these two ideal numbers also has the character of a prime number, that is, it does not divide a product $\omega\omega'$ unless it divides at least one of the factors ω, ω'. In fact if we put

$$\omega = x + y\theta, \quad \omega' = x' + y'\theta, \quad \omega'' = \omega\omega' = x'' + y''\theta,$$

then we have

$$x'' = xx' - 5yy', \quad y'' = xy' + yx',$$

and hence

$$x'' \pm y'' \equiv (x \pm y)(x' \pm y') \quad (\text{mod } 3),$$

which immediately justifies our assertion, bearing in mind the congruences (β_1), (β_2) above. Because of this, the number 3 ought to be considered, from a certain point of view, as the product of the two different ideal prime numbers β_1, β_2.

Moreover, since each of the ideal prime numbers β_1, β_2 is different (in the sense indicated above) from the ideal prime number α introduced above, then in view of the fact that 2 behaves like the square of α and $1 + \theta$ is divisible by α and β_1, and $1 - \theta$ by α and β_2, we ought to conclude from the equation $2 \cdot 3 = (1 + \theta)(1 - \theta)$ that $1 + \theta$ behaves like the product of α and β_1, and $1 - \theta$ like the product of α and β_2. In fact this *presumption* is plainly confirmed: each number $\omega = x + y\theta$ divisible by $1 + \theta$ is in fact divisible by α and β_1, because

$$x + y\theta = (1 + \theta)(x' + y'\theta)$$

implies

$$x = x' - 5y', \quad y = x' + y',$$

and consequently

$$x \equiv y \;(\text{mod } 2), \qquad x \equiv y \;(\text{mod } 3).$$

Conversely, each number $\omega = x + y\theta$ divisible by α and β_1, that is, satisfying the two preceding congruences, is also divisible by $1 + \theta$, because we have $y = x + 6y'$ and consequently

$$x + y\theta = (1 + \theta)(x + 5y' + y'\theta).$$

We can now also introduce the *powers* of the ideal prime numbers β_1, β_2, as we have done above for powers of the ideal number α. By analogy with the theory of rational numbers, we define divisibility of an arbitrary number ω by β_1^n or β_2^n by the respective congruences

(β_1^n) $\qquad\qquad\qquad \omega(1 - \theta)^n \equiv 0 \;(\text{mod } 3^n),$

(β_2^n) $\qquad\qquad\qquad \omega(1 + \theta)^n \equiv 0 \;(\text{mod } 3^n),$

and this yields a series of theorems which agree perfectly with those of the theory of rational numbers. We treat the ideal prime numbers γ_1, γ_2 in the same way.

§*10. Laws of divisibility in the domain* o

By similar study of the whole domain o of numbers $\omega = x + y\theta$ we find the following results:

    1. All the positive rational primes $\equiv 11, 13, 17, 19 \;(\text{mod } 20)$ behave like actual prime numbers.

    2. The number θ with square -5 has the character of a prime number. The number 2 behaves like the square of an ideal prime number α.

    3. Each positive rational prime $\equiv 1, 9 \;(\text{mod } 20)$ can be decomposed into two different factors, which really exist and have the character of primes.

4. Each positive rational prime $\equiv 3, 7 \pmod{20}$ behaves like the product of two different ideal prime numbers.

5. Each actual number ω different from zero and ± 1 is either one of the numbers mentioned above as having the character of a prime, or else it behaves in all questions of divisibility as a unique product of actual or ideal prime factors.

However, to arrive at this result and to become completely certain that the general laws of divisibility governing the domain of rational numbers extend to our domain $\mathfrak{o}$ with the help of the ideal numbers we have introduced,† it is necessary, as we shall soon see when we attempt a rigorous derivation, to make a very deep investigation, even supposing knowledge of the theory of quadratic residues and binary quadratic forms (a theory which, conversely, can be derived with great facility from the general theory of algebraic integers). We can indeed reach the proposed goal with all rigour; however, as we have remarked in the Introduction, the greatest circumspection is necessary to avoid being led to premature conclusions. In particular, the notion of *product* of arbitrary factors, actual or ideal, cannot be exactly defined without going into minute detail. Because of these difficulties, it has seemed desirable to replace the ideal number of Kummer, which is never defined in its own right, but only as a divisor of actual numbers ω in the domain $\mathfrak{o}$, by a *noun* for something which actually exists, and this can be done in several ways.

One can, for example (and if I am not mistaken, this is the way chosen by Kronecker in his researches), replace the ideal numbers by actual algebraic numbers, not from the domain $\mathfrak{o}$, but rather *adjoined* to this domain in the sense of Galois. Indeed, if we put

$$\beta_1 = \sqrt{-2 + \theta}, \quad \beta_2 = \sqrt{-2 - \theta},$$

and if we choose the square roots so that $\beta_1 \beta_2 = 3$ then we have

$$\theta^2 = -5, \quad \beta_1^2 = -2 + \theta, \quad \beta_2^2 = -2 - \theta,$$

$$\beta_1 \beta_2 = 3, \quad \theta \beta_1 = -2\beta_1 - 3\beta_2, \quad \theta \beta_2 = 3\beta_1 + 2\beta_2,$$

whence it follows that the quadrinomial numbers

$$x + y\theta + z_1 \beta_1 + z_2 \beta_2,$$

---

† To some people it seems evident *a priori* that the establishment of this harmony with the theory of rational numbers can be *imposed*, whatever happens, by the introduction of ideal numbers. However the example, given above, of the irregular role of the number 2 in the domain of numbers $x + y\sqrt{-3}$ suffices to dispel this illusion.

where x, y, z_1, z_2 are rational integers, are closed under addition, subtraction and multiplication. The domain o' of these numbers contains the domain o, and all the ideal numbers needed for the latter can be replaced by actual numbers of the new domain o'. For example, by putting

$$\alpha = \beta_1 + \beta_2, \quad \gamma_1 = 2\beta_1 + \beta_2, \quad \gamma_2 = \beta_1 + 2\beta_2$$

all the equations (4) of §7 are satisfied. Likewise, the two ideal prime factors of the number 23 in the domain o are replaced by the two actual numbers $2\beta_1 - \beta_2$ and $-\beta_1 + 2\beta_2$ of the domain o', and it is the same for all the ideal numbers of the domain o.

Although this way is capable of leading to our goal, it does not seem to me as simple as desirable, because one is forced to pass from the given domain o to a more complicated domain o'. It is also easy to see that the choice of the new domain o' is highly arbitrary. In the Introduction I have explained in detail the train of thought that led me to build this theory on quite a different basis, namely on the notion of *ideal*, and it would be superfluous to come back to it here; hence I shall confine myself to illustrating the notion by an example.

§11. *Ideals in the domain* o

The condition for a number $\omega = x + y\theta$ to be divisible by the ideal prime number α is that $x \equiv y \pmod 2$, by §8. Thus to obtain the system $\mathfrak{a}$ of all numbers ω divisible by α we put $x = y + 2z$, where y and z are arbitrary rational integers. The system $\mathfrak{a}$ therefore consists of all numbers of the form $2z + (1+\theta)y$, that is, $\mathfrak{a}$ is a *finitely generated module* with basis consisting of the two independent numbers 2 and $1 + \theta$, and consequently

$$\mathfrak{a} = [2, 1 + \theta].$$

Similarly letting $\mathfrak{b}_1$, $\mathfrak{b}_2$, $\mathfrak{c}_1$, $\mathfrak{c}_2$ denote the systems of all numbers ω divisible by β_1, β_2, γ_1, γ_2 respectively, we conclude from the corresponding congruences in §9 that

$$\mathfrak{b}_1 = [3, 1 + \theta], \quad \mathfrak{b}_2 = [3, 1 - \theta],$$
$$\mathfrak{c}_1 = [7, 3 + \theta], \quad \mathfrak{c}_2 = [7, 3 - \theta].$$

If we now let $\mathfrak{m}$ denote any one of these five systems, then $\mathfrak{m}$ enjoys the following properties.

I. The sum and difference of any two numbers in m are also in m.

II. Each product of a number in the system m by a number in the system o is a number in the system m.

The first property, characteristic of each module, is evident. To establish the second property for a system m whose basis consists of the numbers μ, μ' it evidently suffices to show that the two products $\theta\mu$, $\theta\mu'$ belong to the same system. For the system a this follows from the two equations

$$2\theta = -1 \cdot 2 + 2(1 + \theta), \quad (1 + \theta)\theta = -3 \cdot 2 + (1 + \theta),$$

and it is just the same for the other systems. But these two properties can also be established without these verifications, by appealing to the fact that each of the five systems m is the set of all numbers ω in the domain o satisfying a congruence of the form

$$\nu\omega \equiv 0 \pmod{\mu},$$

where μ, ν are two given numbers in the domain o.

We now call *any* system m of numbers in domain o enjoying properties I and II an *ideal*, and we begin by posing the problem of finding the general *form* of all ideals. Excluding the singular case where m consists of the single number zero, we choose an arbitrary nonzero number μ in m. Then if μ' denotes the conjugate number, the norm $N(\mu) = \mu\mu'$, and hence the product $\theta N(\mu)$ also belongs to the ideal m by virtue of II. Thus all the numbers in the module o = $[1, \theta]$, when multiplied by the nonzero rational number $N(\mu)$, become numbers in the module m, which is at the same time a *multiple* of o. But it is a consequence of (§3,2) that m is a finitely generated module, of the form $[k, l + m\theta]$ where k, l, m are rational integers, among which k and m can be chosen *positive*. Since m already has property I, as a module, the question is what follows from property II, which says that the two products $k\theta$ and $(l + m\theta)\theta$ belong to the system m. The necessary and sufficient conditions for this, as one sees without difficulty, are that m divide k and l and that the rational integers a, b appearing in the expression

$$\mathfrak{m} = [ma, m(b + \theta)]$$

also satisfy the congruence

$$b^2 \equiv -5 \pmod{a}.$$

If we replace b by any number $\equiv b \pmod{a}$ then the ideal m is unchanged.

The five ideals above, $\mathfrak{a}$, $\mathfrak{b}_1$, $\mathfrak{b}_2$, $\mathfrak{c}_1$, $\mathfrak{c}_2$ are evidently of this form, since $(b + \theta)$ can also be replaced by $-(b + \theta)$.

The set of all numbers conjugate to the numbers in an ideal $\mathfrak{m}$ is evidently also an ideal

$$\mathfrak{m}_1 = [ma, m(-b + \theta)].$$

Two such ideals $\mathfrak{m}$, $\mathfrak{m}_1$ may be called *conjugate* ideals.

Let μ be any number in the domain o. The system $[\mu, \mu\theta]$ of all numbers divisible by μ forms an ideal which we call a *principal ideal*,† and which we denote by $\mathfrak{o}(\mu)$ or $\mathfrak{o}\mu$. It is easy to give it the above form $[ma, m(b + \theta)]$; m is the greatest rational integer that divides $\mu = m(u + v\theta)$ and we have, moreover,

$$a = \frac{N(\mu)}{\mu^2}, \quad vb \equiv u \pmod{a}.$$

Thus we find, for example,

$$\mathfrak{o}(\pm 1) = \mathfrak{o} = [1, \theta],$$

and

$$\mathfrak{o}(2) = [2, 2\theta], \quad \mathfrak{o}(3) = [3, 3\theta], \quad \mathfrak{o}(7) = [7, 7\theta],$$
$$\mathfrak{o}(1 \pm \theta) = [6, \pm 1 + \theta], \quad \mathfrak{o}(3 \pm \theta) = [14, \pm 3 + \theta],$$
$$\mathfrak{o}(-2 \pm \theta) = [9, \mp 2 + \theta], \quad \mathfrak{o}(2 \pm 3\theta) = [49, \pm 17 + \theta],$$
$$\mathfrak{o}(-1 \pm 2\theta) = [21, \pm 10 + \theta], \quad \mathfrak{o}(4 \pm \theta) = [21, \pm 4 + \theta].$$

Since all ideals are also modules, we say (following §2,1) that two numbers ω, ω' are *congruent* modulo the ideal $\mathfrak{m}$, and put $\omega \equiv \omega' \pmod{\mathfrak{m}}$ when $\omega - \omega'$ is a number in $\mathfrak{m}$. The *norm* $N(\mathfrak{m})$ of the ideal $\mathfrak{m} = [ma, m(b + \theta)]$ is the number

$$(\mathfrak{o}, \mathfrak{m}) = m^2 a$$

of *classes* into which the domain o is partitioned modulo the module $\mathfrak{m}$ (§4,4). If $\mathfrak{m}$ is a principal ideal $\mathfrak{o}\mu$ then the preceding congruence will be equivalent to $\omega \equiv \omega' \pmod{\mu}$ and we shall have

$$N(\mathfrak{m}) = N(\mu).$$

The norm of any number $m(ax + (b+\theta)y)$ in the ideal $\mathfrak{m} = [ma, m(b+\theta)]$

---

† If we extend the definition of ideal to the domain o of rational integers, or to the complex integers of Gauss, or to any of the five domains o considered in §7, then one easily sees that every ideal is a principal ideal. It is also evident that, in the domain of rational integers, property II is already contained in property I.

is equal to the product of $N(\mathfrak{m}) = m^2 a$ with the binary quadratic form $ax^2 + 2bxy + cy^2$ whose determinant, according to the definition of Gauss, is $b^2 - ac = -5$.‡

§12. *Divisibility and multiplication of ideals in* o

I shall now show how the theory of numbers $\omega = x + y\theta$ in the domain $\mathfrak{m}$ can be based on the notion of ideal. For the sake of brevity, however, I shall be obliged to leave certain easy calculations to the reader.

Just as in the theory of modules (§1,2), we say that an ideal $\mathfrak{m}''$ is *divisible* by an ideal $\mathfrak{m}$ when all numbers in $\mathfrak{m}''$ belong to $\mathfrak{m}$. It follows that a principal ideal $\mathfrak{o}\mu''$ is divisible by a principal ideal $\mathfrak{o}\mu$ if and only if the number μ'' is divisible by the number μ. Thus the theory of divisibility of numbers *is contained in* the theory of divisibility of ideals. One sees immediately that the necessary and sufficient conditions for the ideal $\mathfrak{m}'' = [m''a'', m''(b'' + \theta)]$ to be divisible by the ideal $\mathfrak{m} = [ma, m(b + \theta)]$ are the three congruences

$$m''a \equiv m''a'' \equiv m''(b'' - b) \equiv 0 \pmod{ma}.$$

The definition of *multiplication* of ideals is the following: if μ runs through the numbers in the ideal $\mathfrak{m}$, and μ' through the numbers in the ideal $\mathfrak{m}'$, then all the products $\mu\mu'$ and their sums form an ideal $\mathfrak{m}''$ called the *product*† of the factors $\mathfrak{m}$, $\mathfrak{m}'$ and denoted by $\mathfrak{mm}'$. We evidently have

$$\mathfrak{om} = \mathfrak{m}, \quad \mathfrak{mm}' = \mathfrak{m}'\mathfrak{m}, \quad (\mathfrak{mm}')\mathfrak{n} = \mathfrak{m}(\mathfrak{m}'\mathfrak{n}),$$

whence it follows that products of any number of ideals satisfy the same theorems as products of numbers.‡ Moreover, it is clear that the product of two principal ideals $\mathfrak{o}\mu$ and $\mathfrak{o}\mu'$ is the principal ideal $\mathfrak{o}(\mu\mu')$.

Now given two ideals

$$\mathfrak{m} = [ma, m(b + \theta)], \quad \mathfrak{m}' = [m'a', m'(b' + \theta)],$$

we derive their product

$$\mathfrak{m}'' = \mathfrak{mm}' = [m''a'', m''(b'' + \theta)]$$

‡ The general theory of forms is nevertheless simplified a little when we also admit the forms $Ax^2 + Bxy + Cy^2$, where B is odd, and if we always understand the determinant of the form to be the number $B^2 - 4AC$.
† The same definition also applies for multiplication of two *modules*.
‡ See Dirichlet's *Vorlesungen über Zahlentheorie*, §2.

with the help of the methods indicated in the first chapter (§4, 5 and 6).
It is clear from the definition that the product $\mathfrak{m}\mathfrak{m}'$ is a finitely generated
module with basis consisting of the *four* products

$$mm'aa', \quad mm'a(b'+\theta), \quad mm'a'(b+\theta),$$
$$mm'(b+\theta)(b'+\theta) = mm'[bb' - 5 + (b+b')\theta],$$

of which only *two* are independent. Thus for the ideals considered above,
for example

$$\mathfrak{b}_1 = [3, 1+\theta], \quad \mathfrak{c}_2 = [7, 3-\theta],$$

we find the product

$$\mathfrak{b}_1\mathfrak{c}_2 = [21, 9 - 3\theta, 7 + 7\theta, 8 + 2\theta].$$

This module is derived from the one considered at the end of the first
chapter (§4,6), and by setting $\omega_1 = 1$, $\omega_2 = \theta$ we conclude

$$\mathfrak{b}_1\mathfrak{c}_2 = [21, -17+\theta] = [21, 4+\theta] = \mathfrak{o}(4+\theta).$$

In the same way we obtain all the following results, completely analogous
to the hypothetical equations (4) of §7:

$$\mathfrak{o}(2) = \mathfrak{a}^2, \quad \mathfrak{o}(3) = \mathfrak{b}_1\mathfrak{b}_2, \quad \mathfrak{o}(7) = \mathfrak{c}_1\mathfrak{c}_2;$$
$$\mathfrak{o}(-2+\theta) = \mathfrak{b}_1^2, \quad \mathfrak{o}(-2-\theta) = \mathfrak{b}_2^2;$$
$$\mathfrak{o}(2+3\theta) = \mathfrak{c}_1^2, \quad \mathfrak{o}(2-3\theta) = \mathfrak{c}_2^2;$$
$$\mathfrak{o}(1+\theta) = \mathfrak{a}\mathfrak{b}_1, \quad \mathfrak{o}(1-\theta) = \mathfrak{a}\mathfrak{b}_2;$$
$$\mathfrak{o}(3+\theta) = \mathfrak{a}\mathfrak{c}_1, \quad \mathfrak{o}(3-\theta) = \mathfrak{a}\mathfrak{c}_2;$$
$$\mathfrak{o}(-1+2\theta) = \mathfrak{b}_1\mathfrak{c}_1, \quad \mathfrak{o}(-1-2\theta) = \mathfrak{b}_2\mathfrak{c}_2;$$
$$\mathfrak{o}(4+\theta) = \mathfrak{b}_1\mathfrak{c}_2, \quad \mathfrak{o}(4-\theta) = \mathfrak{b}_2\mathfrak{c}_1.$$

To effect the multiplication of two ideals; $\mathfrak{m}$, $\mathfrak{m}'$ *in general* it is necessary
to transform the basis of the four numbers above into one consisting of
only two numbers $m''a'', m''(b''+\theta)$. One arrives at this (by virtue of
§4) via four equations of the form

$$mm'aa' = pm''a'' + qm''(b''+\theta),$$
$$mm'a(b'+\theta) = p'm''a'' + q'm''(b''+\theta),$$
$$mm'a'(b+\theta) = p''m''a'' + q''m''(b''+\theta),$$
$$mm'[bb' - 5 + (b+b')\theta] = p'''m''a'' + q'''m''(b''+\theta),$$

where $p, p', \ldots, q'''$ denote eight rational integers chosen so that the six determinants formed from them,

$$P = pq' - qp', \quad Q = pq'' - qp'', \quad R = pq''' - qp''',$$
$$U = p''q''' - q''p''', \quad T = p'q''' - q'p''', \quad S = p'q'' - q'p''$$

have no common divisor. From the four preceding equations, each of
which decomposes into two, we now conclude without difficulty that
these six determinants are respectively proportional to the six numbers

$$a, \quad a', \quad b' + b,$$
$$c, \quad c', \quad b' - b,$$

where c and c' are determined by the equations

$$bb - ac = b'b' - a'c' = -5.$$

But, since these six numbers admit no common divisor,† they must
precisely coincide with the six determinants. It follows, since $q = 0$
and q', q'', q''' can have no common divisor, that we can determine the
product $\mathfrak{m}'' = \mathfrak{m}\mathfrak{m}'$ of two given factors $\mathfrak{m}$, $\mathfrak{m}'$ as follows. Let p be the
greatest common (positive) divisor of the three given numbers

$$a = pq', \quad a' = pq'', \quad b + b' = pq'''.$$

We have

$$\mathfrak{m}'' = p\mathfrak{m}\mathfrak{m}', \quad a'' = \frac{aa'}{p^2} = q'q'',$$

and b'' is determined by the congruences

$$q'b'' \equiv q'b', \quad q''b'' \equiv q''b, \quad q'''b'' \equiv \frac{bb' - 5}{p} \quad (\text{mod } a'').$$

At the same time we have $b''b'' \equiv -5 \pmod{a''}$, that is

$$b''b'' - a''c'' = -5,$$

where c'' is a rational integer and, to use a terminology employed by
Gauss,‡ the binary quadratic form (a'', b'', c'') is *composed* from the two
forms (a, b, c) and (a', b', c').

The values of m'', a'' yield $m''^2 a'' = m^2 a \cdot m'^2 a'$, and hence the theorem

$$N(\mathfrak{m}\mathfrak{m}') = N(\mathfrak{m})N(\mathfrak{m}').$$

† This will not always be so in the domain of numbers $x + y\sqrt{-3}$.
‡ *Disquisitiones Arithmeticae*, art. 235, 242.

It is also necessary to note the special case where $\mathfrak{m}'$ is the ideal $\mathfrak{m}_1$ conjugate to $\mathfrak{m}$. The preceding formulas then yield the immediate result

$$\mathfrak{m}\mathfrak{m}_1 = \mathfrak{o}N(\mathfrak{m}).$$

The two notions of *divisibility* and *multiplication* of ideals are now connected in the following manner. The product $\mathfrak{m}\mathfrak{m}'$ is divisible by both $\mathfrak{m}$ and $\mathfrak{m}'$ since, by property II of ideals, all the products $\mu\mu'$ whose factors belong respectively to $\mathfrak{m}$, $\mathfrak{m}'$ are in both these ideals; the same then holds for the product ideal itself. Conversely, if the ideal $\mathfrak{m}'' = [m''a'', m''(b'' + \theta)]$ is divisible by the ideal $\mathfrak{m} = [m, m(b + \theta)]$ then there is exactly one ideal $\mathfrak{m}'$ such that $\mathfrak{m}\mathfrak{m}' = \mathfrak{m}''$. In fact, if we let $\mathfrak{m}_1$ denote the ideal conjugate to $\mathfrak{m}$ and form the product

$$\mathfrak{m}_1\mathfrak{m}'' = [m'''a', m'''(b' + \theta)]$$

by the preceding rules, then it follows from the three congruences established at the beginning of this § that m''' is divisible by $N(\mathfrak{m}) = m^2a$, and hence that $m''' = m^2am'$, where m' is an integer. Combining this with the preceding theorem that $\mathfrak{m}\mathfrak{m}_1 = \mathfrak{o}(m^2a)$, we easily conclude that the ideal $\mathfrak{m}' = [m'a', m'(b' + \theta)]$, and it alone, satisfies the condition $\mathfrak{m}\mathfrak{m}' = \mathfrak{m}''$. At the same time it follows that the equation $\mathfrak{m}\mathfrak{m}' = \mathfrak{m}\mathfrak{m}'''$ always implies $\mathfrak{m}' = \mathfrak{m}'''$.

Now to arrive at the conclusion of this theory, it only remains to introduce the following notion. An ideal $\mathfrak{p}$, different from $\mathfrak{o}$ and divisible by no ideals other than $\mathfrak{o}$ and $\mathfrak{p}$, will be called a *prime ideal*. If η is a particular number, then the system $\mathfrak{r}$ of *all* roots ρ of the congruence $\eta\rho \equiv 0 \pmod{\mathfrak{p}}$ forms an ideal, because it has properties I and II. This ideal $\mathfrak{r}$ is a divisor of $\mathfrak{p}$, because all numbers in $\mathfrak{p}$ are also roots of this congruence. Thus, if $\mathfrak{p}$ is a prime ideal, then $\mathfrak{r}$ must be $\mathfrak{o}$ or $\mathfrak{p}$. If the given number η is not in $\mathfrak{p}$ then the number 1, in $\mathfrak{o}$, will not be a root of the congruence, and hence in this case $\mathfrak{r}$ will be neither $\mathfrak{o}$ nor $\mathfrak{p}$. That is, all the roots ρ must be in $\mathfrak{p}$. Thus we have established the following theorem:†

"A product $\eta\rho$ of two numbers η, ρ is not in a prime ideal $\mathfrak{p}$ unless at least one of the factors is in $\mathfrak{p}$."

And this immediately yields the theorem:

---

† This theorem leads easily to the determination of all prime ideals contained in $\mathfrak{o}$, and they correspond exactly to the prime numbers, actual and ideal, enumerated in §10.

"If neither of the ideals $\mathfrak{m}$, $\mathfrak{m}'$ is divisible by the prime ideal $\mathfrak{p}$ then their product will also not be divisible by $\mathfrak{p}$."

Because if $\mathfrak{m}$, $\mathfrak{m}'$ respectively include numbers μ, μ' not in $\mathfrak{p}$, then $\mathfrak{m}\mathfrak{m}'$ includes the number $\mu\mu'$ not in $\mathfrak{p}$.

Combining the theorem just proved with the preceding theorems on the connection between divisibility and multiplication of ideals, and bearing in mind that $\mathfrak{o}$ is the only ideal with norm 1, we arrive, by exactly the same reasoning† as in the theory of rational numbers, at the following theorem: *"Each ideal different from $\mathfrak{o}$ is either a prime ideal or else uniquely expressible as a product of finitely many prime ideals."* It follows immediately from this theorem that an ideal $\mathfrak{m}''$ is divisible by an ideal $\mathfrak{m}$ if and only if all the powers of prime ideals that divide $\mathfrak{m}$ also divide $\mathfrak{m}''$. If $\mathfrak{m} = \mathfrak{o}\mu$ and $\mathfrak{m}'' = \mathfrak{o}\mu''$ are principal ideals then the same criterion also decides the divisibility of the *number* μ'' by the *number* μ. Thus the theory of divisibility of numbers in the domain $\mathfrak{o}$ is restored to firm and simple laws.

All this theory can be applied almost word for word to any domain $\mathfrak{o}$ consisting of *all* the integers of a quadratic field Ω, when the notion of *integer* is defined as in the Introduction.‡ However, even though this approach to the theory leaves nothing to be desired in the way of rigour, it is not at all what I propose to carry out. One notices, in fact, that the proofs of the most important propositions depend upon the representation of an ideal by the *expression* $[ma, m(b + \theta)]$ and on the effective realisation of multiplication, that is, on a *calculus* which coincides with the composition of binary quadratic forms given by Gauss. If we want to treat fields Ω of arbitrary degree in the same way, then we shall run into great difficulties, perhaps insurmountable ones. Even if there were such a theory, based on calculation, it still would not be of the highest degree of perfection, in my opinion. It is preferable, as in the modern theory of functions, to seek proofs based immediately on fundamental characteristics, rather than on calculation, and indeed to construct the theory in such a way that it is able to predict the results of calculation (for example the composition of decomposable forms of all degrees). Such is the goal I shall pursue in the chapters of this memoir that follow.

---

† See Dirichlet's *Vorlesungen über Zahlentheorie*, §8.

‡ The domain, mentioned above, of numbers $x + y\sqrt{-3}$ where x, y are rational integers is *not* a domain of this nature. However, it forms only a *part* of the domain $\mathfrak{o}$ of all the numbers $x + y\rho$, where ρ is a root of the equation $\rho^2 + \rho + 1 = 0$.

3

General properties of algebraic integers

In this chapter we first consider the domain of all algebraic integers. Then we introduce the notion of a field Ω of finite degree, and determine the constitution of the domain $\mathfrak{o}$ of integers of the field Ω.

§13. The domain of all algebraic integers

A real or complex number θ is called *algebraic* when it satisfies an equation

$$\theta^n + a_1\theta^{n-1} + a_2\theta^{n-2} + \cdots + a_{n-1}\theta + a_n = 0$$

of finite degree n with rational coefficients $a_1, a_2, \ldots, a_{n-1}, a_n$. If the coefficients of this equation are rational *integers*, that is, numbers from the sequence $0, \pm 1, \pm 2, \ldots$, then θ is called an *algebraic integer*, or simply an *integer*. It is clear that the rational integers are also algebraic integers and, conversely, if a rational number θ is at the same time an algebraic integer then, by virtue of a known theorem, it will also be one of the rational integers $0, \pm 1, \pm 2, \ldots$. From the definition of integers we easily derive the following propositions:

1. The integers are closed under addition, subtraction and multiplication, that is, the sum, difference and product of any two integers α, β are also integers.

Proof. By hypothesis, there are two equations of the form

$$\phi(\alpha) = \alpha^a + p_1\alpha^{a-1} + \cdots + p_{a-1}\alpha + p_a = 0,$$
$$\psi(\beta) = \beta^b + q_1\beta^{b-1} + \cdots + q_{b-1}\beta + q_b = 0,$$

in which all the coefficients p, q are rational integers. We now put $ab = n$

and let

$$\omega_1, \omega_2, \ldots, \omega_n$$

denote the n products $\alpha^{a'}\beta^{b'}$ formed from one of the a numbers

$$1, \alpha, \alpha^2, \ldots, \alpha^{a-1}$$

and one of the b numbers

$$1, \beta, \beta^2, \ldots, \beta^{b-1}.$$

If ω now represents one of the three numbers $\alpha + \beta$, $\alpha - \beta$, $\alpha\beta$ then we easily see that each of the n products $\omega\omega_1, \omega\omega_2, \ldots, \omega\omega_n$ can be reduced immediately, with the help of the equations $\phi(\alpha) = 0$, $\psi(\beta) = 0$, to the form

$$k_1\omega_1 + k_2\omega_2 + \cdots + k_n\omega_n,$$

where $k_1, k_2, \ldots, k_n$ are rational *integers*. We then have n equations of the form

$$\omega\omega_1 = k_1'\omega_1 + k_2'\omega_2 + \cdots + k_n'\omega_n,$$
$$\omega\omega_2 = k_1''\omega_1 + k_2''\omega_2 + \cdots + k_n''\omega_n,$$
$$\ldots \ldots \ldots \ldots \ldots \ldots \ldots \ldots \ldots \ldots \ldots \ldots$$
$$\omega\omega_n = k_1^{(n)}\omega_1 + k_2^{(n)}\omega_2 + \cdots + k_n^{(n)}\omega_n,$$

all coefficients k of which are rational integers. Now by elimination of the n numbers $\omega_1, \omega_2, \ldots, \omega_n$, which include the nonzero number 1, we derive the equation

$$\begin{vmatrix} k_1' - \omega & k_2' & \cdots & k_n' \\ k_1'' & k_2'' - \omega & \cdots & k_n'' \\ \cdots & \cdots\cdots & \cdots & \cdots \\ k_1^{(n)} & k_2^{(n)} & \cdots & k_n^{(n)} - \omega \end{vmatrix} = 0$$

which is evidently of the form

$$\omega^n + e_1\omega^{n-1} + \cdots + e_{n-1}\omega + e_n = 0,$$

where the n coefficients e are formed from the numbers k by addition, subtraction and multiplication, and hence are rational integers. Thus ω, and consequently each of the three numbers $\alpha + \beta$, $\alpha - \beta$, $\alpha\beta$, is an integer.   Q.E.D.

2. Each root ω of an equation of the form

$$F(\omega) = \omega^m + \alpha\omega^{m-1} + \beta\omega^{m-2} + \cdots + \epsilon = 0,$$

where the coefficient of the highest degree term is unity and the others $\alpha, \beta, \ldots, \epsilon$ are integers, is likewise an integer.

Proof. By hypothesis, the coefficients $\alpha, \beta, \ldots, \epsilon$ are roots of equations

$$\phi(\alpha) = \alpha^a + p_1 \alpha^{a-1} + \ldots + p_a = 0,$$
$$\psi(\beta) = \beta^b + q_1 \beta^{b-1} + \ldots + q_b = 0,$$
$$\ldots\ldots\ldots\ldots\ldots\ldots\ldots\ldots\ldots\ldots\ldots$$
$$\chi(\epsilon) = \epsilon^e + s_1 \epsilon^{e-1} + \ldots + s_e = 0,$$

where all the coefficients $p, q, \ldots, s$ are rational integers. Now if we put $n = mab \cdots e$ and let $\omega_1, \omega_2, \ldots, \omega_n$ denote all the n products of the form

$$\omega^{m'} \alpha^{a'} \beta^{b'} \cdots \epsilon^{e'},$$

where the exponents are rational integers satisfying

$$0 \leq m' < m, \quad 0 \leq a' < a, \quad 0 \leq b' < b, \quad \ldots, \quad 0 \leq e' < e,$$

then it is easily seen that, with the help of the equations $F(\omega) = 0$, $\phi(\alpha) = 0$, $\psi(\beta) = 0$, $\ldots$, $\chi(\epsilon) = 0$, all the products $\omega \omega_1, \omega \omega_2, \ldots, \omega \omega_n$ reduce immediately to the form

$$k_1 \omega_1 + k_2 \omega_2 + \cdots + k_n \omega_n,$$

where $k_1, k_2, \ldots, k_n$ are rational integers. It then follows, as in the previous proof, that ω is an integer.   Q.E.D.

It follows from the latter theorem that, for example, if α is any integer and r, s are positive rational integers then $\sqrt[s]{\alpha^r}$ is also an integer.

§14. *Divisibility of integers*

We say that an integer α is *divisible* by an integer β when $\alpha = \beta\gamma$ for some integer γ. The same thing is expressed by saying that α is a multiple of β, or that β divides α, or that β is a factor or divisor of α. It follows from this definition and Theorem 1 of §13 that we have the following two elementary propositions, mentioned in the Introduction.

1. If α, α' are divisible by μ then $\alpha + \alpha'$, $\alpha - \alpha'$ are also divisible by μ;

2. If α' is divisible by α and α is divisible by μ then α' is also divisible by μ.

However, it is necessary to pay particular attention to *units*, that is, to the integers that divide *all* integers. A unit ϵ must therefore divide the number 1, and conversely it is evident that every divisor ϵ of the number 1 is a unit, because every integer is divisible by 1, and hence (by virtue of proposition 2 above) also by ϵ. At the same time we see that each product and quotient of two units is itself a unit.

If each of two nonzero integers α, α' is divisible by the other then we have $\alpha' = \alpha\epsilon$, where ϵ is a unit. Conversely, if ϵ is a unit then each of the two integers α and $\alpha' = \alpha\epsilon$ is divisible by the other. We call two numbers α, α' *associates* if they are related in this way, and it is clear that any associates of a third number are associates of each other. In all questions concerning just divisibility, associates behave as a single number. Indeed, if α is divisible by β, then every associate of α is divisible by every associate of β.

A deeper investigation will enable us to see that two nonzero integers α, β have a *greatest common divisor*, which can be put in the form $\alpha\alpha' + \beta\beta'$ where α' and β' are integers. This important theorem is not at all easy to prove with the help of the principles developed thus far, but we shall later (§30) be able to derive it very simply from the theory of ideals. I shall therefore end these preliminary considerations of the domain of *all* integers with the remark that there are absolutely no numbers in this domain with the character of *prime numbers*. Because, if α is a nonzero integer, and not a unit, then we can decompose it in infinitely many ways into factors which are integers but not units. For example, we have $\alpha = \sqrt{\alpha} \cdot \sqrt{\alpha}$ and also $\alpha = \beta_1\beta_2$ where β_1, β_2 are the two roots β of the equation $\beta^2 - \beta + \alpha = 0$. It follows from Theorem 2 of §13 that $\sqrt{\alpha}$, β_1, β_2 are integers as well as α.

§15. Fields of finite degree

The property of being decomposable in infinitely many ways, whose presence has just been pointed out in the domain of all integers, disappears again when the integers under consideration are confined to a *field of finite degree*. First we have to define the extent and nature of such a field.

Each algebraic number θ, whether an integer or not, evidently satisfies an infinity of different equations with rational coefficients, that is, there is an infinity of polynomial functions $F(t)$ of one variable t which vanish for $t = \theta$ and whose coefficients are rational. However, among all these functions $F(t)$ there is necessarily one $f(t)$ whose degree n is *as small as possible*, and it follows immediately from the well-known method of polynomial division that each of the functions $F(t)$ will be algebraically divisible by this function $f(t)$, and that $f(t)$ will not be divisible by any polynomial of lower degree and rational coefficients. For this reason, the function $f(t)$ and also the equation $f(\theta) = 0$ are called *irreducible*, and it is clear at the same time that the n numbers $1, \theta^1, \theta^2, \ldots, \theta^{n-1}$ form an *irreducible system* (§4,1).

We now consider the set Ω of all numbers ω of the form $\phi(\theta)$, where

$$\phi(t) = x_0 + x_1 t + x_2 t^2 + \cdots + x_{n-1} t^{n-1}$$

is any polynomial in t with rational coefficients $x_0, x_1, x_2, \ldots, x_{n-1}$ and degree $< n$. We first remark that each such number $\omega = \phi(\theta)$ is uniquely expressible in this form, by virtue of the irreducibility of $f(t)$. We then see easily that the numbers ω are closed under *rational operations*, that is, addition, subtraction, multiplication and division. For the first two operations this evidently follows from the common form $\phi(\theta)$ of all the numbers ω, and for multiplication it suffices to remark that each number of the form $\psi(\theta)$, where $\psi(t)$ is a polynomial of *any* degree with rational coefficients, is likewise a number ω, because if we divide $\psi(t)$ by $f(t)$ the remainder will be a function $\phi(t)$ of the kind above, and at the same time we have $\psi(\theta) = \phi(\theta)$. Finally, to treat the case of division it is enough to show that if $\omega = \phi(\theta)$ is nonzero then its reciprocal ω^{-1} also belongs to the system Ω. Well, since $\phi(t)$ has no common divisor with the irreducible function $f(t)$, the method for finding the greatest common divisor of the polynomials $f(t)$, $\phi(t)$ gives, as we know, two polynomials $f_1(t)$, $\phi_1(t)$, with rational coefficients, satisfying the identity

$$f(t)f_1(t) + \phi(t)\phi_1(t) = 1.$$

When $t = \theta$ this gives the result claimed.

I call a system A of numbers a (not all zero) a *field* when the sum, difference, product and quotient of any two numbers in A also belongs to A. The simplest example of a field is the system of rational numbers. It is easy to see that this field is contained in any other field A since, if we choose any nonzero number a in A, it is necessary that the quotient 1 of the numbers a and a belong to A, whence the claim follows, since all the rational numbers are engendered from the number 1 by repeated additions, subtractions, multiplications and divisions.

Our system Ω is a field, by the results just proved for the numbers $\omega = \phi(\theta)$. The rational numbers come from $\phi(\theta)$ by annulling all the coefficients $x_1, x_2, \ldots, x_{n-1}$ which follow x_0. A field Ω obtained from an irreducible equation $f(\theta) = 0$ of degree n in the manner indicated is called a field of *finite degree*,† and the number n is called its *degree*. Such a field Ω includes n independent numbers, for example $1, \theta, \theta^2, \ldots, \theta^{n-1}$, whereas any $n+1$ numbers in the field evidently form a reducible system (§4,1). The latter property, combined with the definition of field, can also serve as the definition of a field Ω of nth degree. However, I shall not go into the proof of this assertion.

Now if we arbitrarily choose n numbers

$$\omega_1 = \phi_1(\theta), \quad \omega_2 = \phi_2(\theta), \quad \ldots, \quad \omega_n = \phi_n(\theta)$$

of the field Ω, these numbers (by §4,2) form an irreducible system if and only if the determinant of their n^2 rational coefficients x is nonzero. In this case we call the system of n numbers $\omega_1, \omega_2, \ldots, \omega_n$ a *basis of the field* Ω. Then it is evident that each number $\omega = \phi(\theta)$ is always expressible, uniquely, in the form

$$\omega = h_1\omega_1 + h_2\omega_2 + \cdots + h_n\omega_n,$$

with rational coefficients $h_1, h_2, \ldots, h_n$. And conversely, all numbers ω of this form are in Ω. The rational coefficients $h_1, h_2, \ldots, h_n$ will be called the *coordinates of the number ω* with respect to this basis.

§16. Conjugate fields

We ordinarily understand *substitution* to be an act by which objects or elements being studied are replaced by corresponding objects or elements, and we say that the old elements are changed, by the substitution, into the new. Now let Ω be *any* field. By an *isomorphism‡ of* Ω we mean a substitution which changes each number

$$\alpha, \quad \beta, \quad \alpha + \beta, \quad \alpha - \beta, \quad \alpha\beta, \quad \alpha/\beta$$

---

† If we understand a divisor of a field A to be any field B whose members all belong to A, then a field of finite degree can also be defined as a field with only a finite number of divisors. The word *divisor* (and the word *multiple*) is used here in a sense directly opposite to that used in speaking of modules and ideals, but this should not result in any confusion.

‡ Dedekind calls it a *permutation*. (Translator's note.)

of Ω into a corresponding number

$$\alpha', \quad \beta', \quad (\alpha + \beta)', \quad (\alpha - \beta)', \quad (\alpha\beta)', \quad (\alpha/\beta)'$$

in such a way that the conditions

(1) $$(\alpha + \beta)' = \alpha' + \beta',$$
(2) $$(\alpha\beta)' = \alpha'\beta',$$

are satisfied and the substitute numbers $\alpha', \beta', \ldots$ are not all zero. We shall see that the set Ω' of the latter numbers forms a new field, and that the isomorphism also satisfies the following conditions:

(3) $$(\alpha - \beta)' = \alpha' - \beta',$$
(4) $$(\alpha/\beta)' = \alpha'/\beta'.$$

Indeed, if we let α', β' be any two numbers in the system Ω', then there will be two numbers α, β in the field Ω changed respectively into α', β'. But since the numbers $\alpha + \beta$, $\alpha\beta$ are likewise in Ω, it follows from (1) and (2) that the numbers $\alpha' + \beta'$, $\alpha'\beta'$ are in Ω'. Thus the numbers of the system Ω' are closed under addition and multiplication. Moreover, since the numbers $\alpha = (\alpha - \beta) + \beta$ and $\alpha - \beta$ are likewise in Ω, it follows from (1) that

$$\alpha' = (\alpha - \beta)' + \beta',$$

which is condition (3). Thus the numbers of the system Ω' are also closed under subtraction. Finally, if β' is nonzero then, by virtue of (1), β will also be nonzero and hence α/β is a number in the field Ω. Since we now have $\alpha = (\alpha/\beta)\beta$, it follows from (2) that we also have $\alpha' = (\alpha/\beta)'\beta'$, which is condition (4). Thus the numbers of the system Ω' are also closed under division, and consequently Ω' is a field. Q.E.D.

We also note that if $\beta' = 0$ then so too is $\beta = 0$; otherwise *every* number α in the field Ω could be put in the form $(\alpha/\beta)\beta$, with the result that $\alpha' = (\alpha/\beta)'\beta' = 0$, whereas we have agreed that the numbers α' in the system Ω' are not all zero. It evidently follows from this, in view of (3), that an isomorphism changes two *different* numbers α, β in the field Ω into two *different* numbers α', β' in the field Ω', so that each number α' in the field Ω' corresponds to a single number α in the field Ω. The correspondence can therefore be reversed in a unique manner, and the substitution that changes each number α' in the field Ω' into the corresponding number α in the field Ω will be an *isomorphism of the field Ω'*, because it satisfies the characteristic conditions (1) and (2).

Each of the two isomorphisms will be called the *inverse* of the other. We also call Ω and Ω' *conjugate fields*, and any two corresponding numbers α, α' will be called *conjugate numbers*. For each field Ω there is evidently an *identity* isomorphism of Ω, which replaces each number in Ω by itself. Thus every field is conjugate to itself. It is also easy to see that two fields conjugate to a third are conjugate to each other. Because if each number α in a field Ω is changed by an isomorphism P into a number α' in a field Ω', and α' is likewise changed by an isomorphism P' into a number α'' in a field Ω'', then it is clear that the substitution changing each α in Ω into the corresponding α'' in Ω'' is likewise an *isomorphism* of the field Ω, and we denote it by PP'. If we let P^{-1} denote the inverse isomorphism of P, then PP^{-1} will be the identity isomorphism of Ω, and Ω'' will be changed into Ω by the isomorphism

$$(PP')^{-1} = P'^{-1}P^{-1}.$$

We have previously remarked that each field includes all the rational numbers, and it is easy to show that each of these is changed into itself by an isomorphism of the field. Because, if we take $\alpha = \beta$ in (4) we get $1' = 1$, and then, since each rational number can be engendered from 1 by a series of rational operations, our proposition follows immediately from properties (1), (2), (3), (4).

Now let θ be any number in the field Ω and let $R(t)$ be any rational function of t with rational coefficients. The number $\omega = R(\theta)$, in the case where the denominator of $R(t)$ does not vanish for $t = \theta$, will also be in Ω, and if θ is changed into the number θ' by an isomorphism of the field, then the number ω will be changed into the number $\omega' = R(\theta')$, since it is formed from the number θ, and the rational coefficients of $R(t)$, by rational operations. It follows immediately that if θ is an algebraic number, and hence satisfies an equation of the form $0 = F(\theta)$ with rational coefficients, we must also have $0 = F(\theta')$. Thus each number θ' conjugate to an algebraic number θ is likewise an algebraic number, and if θ is an algebraic integer, θ' will also be an algebraic integer.

After these general considerations, which apply to *all* fields, we return to our example of a field Ω of finite degree n, and pose the problem of finding *all* the isomorphisms of Ω. Since all numbers ω in such a field are, by §15, of the form $\phi(\theta)$ where θ is a root of an irreducible n^{th} degree equation $0 = f(\theta)$, it follows from the preceding results that an isomorphism of Ω will be completely determined by the choice of a root θ' of the equation $0 = f(\theta')$, because when θ is changed into θ', $\omega = \phi(\theta)$ is changed at the same time into $\omega' = \phi(\theta')$. Conversely, if we choose

θ' to be any root of the equation $0 = f(\theta')$, and replace each number $\omega = \phi(\theta)$ in the field Ω by the corresponding number $\omega' = \phi(\theta')$, then this substitution will really be an isomorphism of Ω, that is, it will satisfy conditions (1) and (2). To show this, we let $\phi_1(t), \phi_2(t), \ldots$ denote any particular functions of the form $\phi(t)$. Now if we have

$$\alpha = \phi_1(\theta), \quad \beta = \phi_2(\theta), \quad \alpha + \beta = \phi_3(\theta), \quad \alpha\beta = \phi_4(\theta),$$

and consequently

$$\alpha' = \phi_1(\theta'), \quad \beta' = \phi_2(\theta'), \quad (\alpha + \beta)' = \phi_3(\theta'), \quad (\alpha\beta)' = \phi_4(\theta'),$$

it follows from the equations

$$\phi_3(\theta) = \phi_1(\theta) + \phi_2(\theta), \quad \phi_4(\theta) = \phi_1(\theta)\phi_2(\theta)$$

and the irreducibility of the function $f(t)$ that we have identically

$$\phi_3(t) = \phi_1(t) + \phi_2(t), \quad \phi_4(t) = \phi_1(t)\phi_2(t) + \phi_5(t)f(t).$$

Taking $t = \theta'$, this gives the equations (1) and (2) we have to show. If we now put

$$f(t) = (t - \theta')(t - \theta'') \cdots (t - \theta^{(n)})$$

then the n roots $\theta', \theta'', \ldots, \theta^{(n)}$ will be different, since the irreducible polynomial $f(t)$ cannot have a common divisor with its derivative $f'(t)$, and each of them will correspond to an isomorphism $P', P'', \ldots, P^{(n)}$ of the field Ω. The isomorphism $P^{(r)}$ changes each number $\omega = \phi(\theta)$ of the field Ω into the conjugate number $\omega^{(r)} = \phi(\theta^{(r)})$ of the conjugate field $\Omega^{(r)}$. To avoid misunderstanding, we point out that the n conjugate fields $\Omega^{(r)}$, although derived from Ω by n *different* isomorphisms, nevertheless may not all be different. If they are all the same, Ω will be called a *Galois* or *normal field*.† The algebraic principles of Galois amount to reducing the study of arbitrary fields of finite degree to the study of normal fields; however, lack of space prevents me from going further into this subject.

§17. Norms and discriminants

The *norm* $N(\omega)$ of a number ω in a field Ω of degree n is the product

$$(1) \qquad\qquad N(\omega) = \omega'\omega'' \cdots \omega^{(n)}$$

---

† Nowadays called a Galois or normal *extension* (of the rationals); however, it would distort Dedekind's meaning to use the term "extension" here. (Translator's note.)

of the n conjugate numbers $\omega', \omega'', \ldots, \omega^{(n)}$ of ω under the isomorphisms $P', P'', \ldots, P^{(n)}$. It does not vanish unless $\omega = 0$. If ω is a rational number, then all the n numbers $\omega^{(r)}$ equal ω, and hence the norm of a rational number is its nth power. If α, β are any two numbers in the field Ω we have $(\alpha\beta)^{(r)} = \alpha^{(r)}\beta^{(r)}$ and consequently

$$(2) \qquad\qquad N(\alpha\beta) = N(\alpha)N(\beta).$$

The *discriminant* $\Delta(\alpha_1, \alpha_2, \ldots, \alpha_n)$ of any n numbers $\alpha_1, \alpha_2, \ldots, \alpha_n$ in the field Ω is the square

$$(3) \qquad\qquad \Delta(\alpha_1, \alpha_2, \ldots, \alpha_n) = \left(\sum \pm \alpha_1' \alpha_2'' \cdots \alpha_n^{(n)}\right)^2$$

of the determinant formed by the n^2 numbers $\alpha_i^{(r)}$. By virtue of a well-known proposition in the theory of determinants we then have the relation

$$(4) \qquad\qquad \Delta(1, \theta, \theta^2, \ldots, \theta^{n-1}) = (-1)^{n(n-1)/2} N(f'(\theta)),$$

and since $f'(\theta)$ cannot be zero, by the irreducibility of the polynomial $f(t)$, it follows that the discriminant (4) has a nonzero value.

Now if the n numbers $\omega_1, \omega_2, \ldots, \omega_n$ form a *basis* of the field Ω (§15), and if

$$\omega = h_1\omega_1 + h_2\omega_2 + \cdots + h_n\omega_n$$

is any number in the field then, since the coordinates $h_1, h_2, \ldots, h_n$ are rational numbers, the isomorphism $P^{(r)}$ will change ω into the number

$$\omega^{(r)} = h_1\omega_1^{(r)} + h_2\omega_2^{(r)} + \cdots + h_n\omega_n^{(r)},$$

from which we conclude that

$$(5) \qquad\qquad \Delta(\alpha_1, \alpha_2, \ldots, \alpha_n) = a^2\Delta(\omega_1, \omega_2, \ldots, \omega_n),$$

where a is the determinant of the n^2 coordinates of the n numbers $\alpha_1, \alpha_2, \ldots, \alpha_n$. This implies, first, that the discriminant of the basis $\omega_1, \omega_2, \ldots, \omega_n$ cannot vanish, otherwise *every* discriminant would vanish, contrary to the fact from above that $\Delta(1, \theta, \theta^2, \ldots, \theta^{n-1})$ is nonzero. At the same time it follows that $\Delta(\alpha_1, \alpha_2, \ldots, \alpha_n)$ will vanish if and only if the numbers $\alpha_1, \alpha_2, \ldots, \alpha_n$ are dependent on each other (§4,2), and hence do not form a basis of Ω.

Since the numbers in a field are closed under multiplication, for any number μ in Ω we can put

$$(6) \quad \begin{cases} \mu\omega_1 = m_{1,1}\omega_1 + m_{2,1}\omega_2 + \cdots + m_{n,1}\omega_n, \\ \mu\omega_2 = m_{1,2}\omega_1 + m_{2,2}\omega_2 + \cdots + m_{n,2}\omega_n, \\ \dots\dots\dots\dots\dots\dots\dots\dots\dots\dots\dots\dots \\ \mu\omega_n = m_{1,n}\omega_1 + m_{2,n}\omega_2 + \cdots + m_{n,n}\omega_n, \end{cases}$$

where the n^2 coordinates $m_{i,i'}$ are rational numbers. Applying the n isomorphisms $P^{(r)}$ yields n^2 new numbers of the form

$$\mu^{(r)}\omega_i^{(r)} = m_{1,i}\omega_1^{(r)} + m_{2,i}\omega_2^{(r)} + \cdots + m_{n,i}\omega_n^{(r)}$$

and, since their determinant is

$$N(\mu)\sum \pm\omega_1'\omega_2''\cdots\omega_n^{(n)} = \sum \pm m_{1,1}m_{2,2}\cdots m_{n,n}\sum \pm\omega_1'\omega_2''\cdots\omega_n^{(n)},$$

we conclude that

$$(7) \qquad\qquad N(\mu) = \sum \pm m_{1,1}m_{2,2}\cdots m_{n,n},$$

because the determinant

$$\sum \pm\omega_1'\omega_2''\cdots\omega_n^{(n)} = \sqrt{\Delta(\omega_1,\omega_2,\dots,\omega_n)}$$

is nonzero.

It follows that every norm is a *rational* number and, by virtue of (4) and (5), so is every discriminant. These two propositions could also have been deduced from the theory of transformations of symmetric functions, but I wish to avoid relying on this.

If we replace the μ in the equations (6) by $\mu - z$, where z is any rational number, then the coordinates $m_{i,i'}$ are unchanged except for the $m_{i,i}$ on the diagonal, each of which is replaced by $m_{i,i} - z$. Theorem 7 is then changed into the equation

$$\begin{vmatrix} m_{1,1} - z & m_{2,1} & \dots & m_{n,1} \\ m_{1,2} & m_{2,2} - z & \dots & m_{n,2} \\ \dots & \dots & \dots & \dots \\ m_{1,n} & m_{2,n} & \dots & m_{n,n} - z \end{vmatrix} = (\mu' - z)(\mu'' - z)\cdots(\mu^{(n)} - z),$$

which, being valid for *every* rational value of z, is necessarily an *identity* in z. At the same time we see that the n numbers $\mu', \mu'', \dots, \mu^{(n)}$ conjugate to the number μ are the set of roots of an nth degree equation whose coefficients are rational numbers.

§18. The integers in a field Ω of finite degree

After these preliminaries, we now pass on to our main objective, the study of the *integers* in the field Ω of degree n, the set of which we denote by $\mathfrak{o}$. Since the sum, difference and product of any two integers are also integers (by §13,1), and in Ω (by §15), the domain $\mathfrak{o}$, including all the *rational* integers, is closed under addition, subtraction and multiplication. However, the first thing is to put all these numbers in a common, simple form. The following considerations lead us to it.

Since each algebraic number ω is a root of an equation of the form

$$c\omega^m + c_1\omega^{m-1} + \cdots + c_{m-1}\omega + c_m = 0,$$

whose coefficients $c, c_1, \ldots, c_{m-1}, c_m$ are rational integers, it follows, multiplying by c^{m-1}, that each such number yields an integer $c\omega$ when multiplied by the nonzero rational integer c. Now if the n numbers $\omega_1, \omega_2, \ldots, \omega_n$ form a basis of the field Ω, we can take nonzero rational numbers $a_1, a_2, \ldots, a_n$ so that the n numbers

$$\alpha_1 = a_1\omega_1, \quad \alpha_2 = a_2\omega_2, \quad \ldots, \quad \alpha_n = a_n\omega_n$$

become *integers*, evidently forming a basis of Ω, since they are independent (by §4,2). It follows that their discriminant $\Delta(\alpha_1, \alpha_2, \ldots, \alpha_n)$ will be a rational number (by §17), and in fact a nonzero *integer* since, by definition, it is formed from the integers $\alpha_i^{(r)}$ by addition, subtraction and multiplication. Moreover, we obtain all numbers ω in the field Ω by letting the coefficients $x_1, x_2, \ldots, x_n$ in the expression

$$\omega = x_1\alpha_1 + x_2\alpha_2 + \cdots + x_n\alpha_n$$

run through all rational values. If we restrict their values to be rational *integers* then we certainly obtain only *integers* ω (§13,1). However, it is likely that not *all* integers in the field Ω will be represented in this way. The situation relates to the following very important theorem:

If there is an integer β of the form

$$\beta = \frac{k_1\alpha_1 + k_2\alpha_2 + \ldots + k_n\alpha_n}{k}$$

where $k, k_1, k_2, \ldots, k_n$ are rational integers without common divisor, then there is a basis of the field Ω consisting of n integers $\beta_1, \beta_2, \ldots, \beta_n$ which satisfy the condition

$$\Delta(\alpha_1, \alpha_2, \ldots, \alpha_n) = k^2 \Delta(\beta_1, \beta_2, \ldots, \beta_n).$$

Proof. Since $\beta, \alpha_1, \alpha_2, \ldots, \alpha_n$ are integers, they form a basis of a module $\mathfrak{b} = [\beta, \alpha_1, \alpha_2, \ldots, \alpha_n]$ which includes only integers from the field Ω. But since only n of these $n + 1$ numbers are independent there are (§4,5) n independent numbers $\beta_1, \beta_2, \ldots, \beta_n$ which form a basis of the same module $\mathfrak{b} = [\beta_1, \beta_2, \ldots, \beta_n]$ and which are therefore integers in the field. We then have $n + 1$ equations of the form

$$\beta = c_1\beta_1 + c_2\beta_2 + \cdots + c_n\beta_n,$$
$$\alpha_1 = c_{1,1}\beta_1 + c_{2,1}\beta_2 + \cdots + c_{n,1}\beta_n,$$
$$\alpha_2 = c_{1,2}\beta_1 + c_{2,2}\beta_2 + \cdots + c_{n,2}\beta_n,$$
$$\cdots\cdots\cdots\cdots\cdots\cdots\cdots\cdots\cdots\cdots\cdots$$
$$\alpha_n = c_{1,n}\beta_1 + c_{2,n}\beta_2 + \cdots + c_{n,n}\beta_n,$$

whose $n(n+1)$ coefficients are rational integers and whose $n+1$ partial $n \times n$ determinants, obtained by suppressing one horizontal line, have *no* common divisor (§4,6). If we put

$$\sum \pm c_{1,1}c_{2,2}\cdots c_{n,n} = c,$$

then we have (§17,(5))

$$\Delta(\alpha_1, \alpha_2, \ldots, \alpha_n) = c^2\Delta(\beta_1, \beta_2, \ldots, \beta_n).$$

Substituting the preceding expressions for $\beta, \alpha_1, \alpha_2, \ldots, \alpha_n$ in the equation $k\beta = k_1\alpha_1 + k_2\alpha_2 + \ldots + k_n\alpha_n$ and observing that $\beta_1, \beta_2, \ldots, \beta_n$ are independent, we find that

$$kc_1 = k_1c_{1,1} + k_2c_{1,2} + \cdots + k_nc_{1,n},$$
$$kc_2 = k_1c_{2,1} + k_2c_{2,2} + \cdots + k_nc_{2,n},$$
$$\cdots\cdots\cdots\cdots\cdots\cdots\cdots\cdots\cdots\cdots\cdots$$
$$kc_n = k_1c_{n,1} + k_2c_{n,2} + \cdots + k_nc_{n,n}.$$

If we now replace the elements

$$c_{1,r}, \quad c_{2,r}, \quad \ldots, \quad c_{n,r}$$

in the rth row of the determinant c by the respective elements

$$c_1, \quad c_2, \quad \ldots, \quad c_n,$$

we conclude from the preceding equations, using a well-known theorem, that the partial determinant obtained has value $\frac{ck_r}{k}$. Thus the $n+1$ quantities

$$\frac{ck}{k}, \quad \frac{ck_1}{k}, \quad \frac{ck_2}{k}, \quad \ldots, \quad \frac{ck_n}{k}$$

are rational integers with no common divisor, and since the same is true of the $n+1$ numbers $k, k_1, k_2, \ldots, k_m$, we necessarily have $c = \pm k$. Q.E.D.

If $k > 1$, so that the integer β is not in the module $\mathfrak{a} = [\alpha_1, \alpha_2, \ldots, \alpha_n]$, then there is a basis of the field consisting of n integers $\beta_1, \beta_2, \ldots, \beta_n$ with discriminant $\Delta(\beta_1, \beta_2, \ldots, \beta_n)$ of absolute value $< \Delta(\alpha_1, \alpha_2, \ldots, \alpha_n)$. Now since the discriminant of every integer basis of the field Ω is a rational nonzero integer, as shown above, there must be a basis $\omega_1, \omega_2, \ldots, \omega_n$ whose discriminant

$$\Delta(\omega_1, \omega_2, \ldots, \omega_n) = D$$

has *minimum* absolute value. And it follows immediately from the preceding results that, relative to such a basis, each integer

$$\omega = h_1\omega_1 + h_2\omega_2 + \cdots + h_n\omega_n$$

in the field Ω must have *integer* coordinates $h_1, h_2, \ldots, h_n$. Moreover, an integer ω is not divisible by a rational integer k unless all its coordinates are divisible by k. Conversely, since each system of integer coordinates $h_1, h_2, \ldots, h_n$ produces an integer ω, *the set* o *of all integers in the field Ω is identical with the finitely generated module* $[\omega_1, \omega_2, \ldots, \omega_n]$ whose basis consists of the n independent integers $\omega_1, \omega_2, \ldots, \omega_n$.

The discriminant D of such a basis is a fundamentally important invariant of the field Ω. For this reason we call it the *fundamental number* or the *discriminant of the field* Ω, and represent it by $\Delta(\Omega)$. In the singular case $n = 1$, Ω is the field of *rational* numbers, and we take its discriminant to be the number $+1$. To clarify the general case we again consider $n = 2$, that is, the case of a *quadratic field*.

Each root θ of an irreducible quadratic equation is of the form

$$\theta = a + b\sqrt{d},$$

where d is a nonsquare rational integer which is also not divisible by a square (except 1). The numbers a, b are rational, and b is nonzero. The set of all the numbers $\phi(\theta)$ in the corresponding quadratic field is evidently the set of all numbers of the form

$$\omega = t + u\sqrt{d},$$

where t, u run through all rational values. Under the nonidentity isomorphism of the field, $\sqrt{d}$ changes into $-\sqrt{d}$, and hence ω changes into the conjugate number

$$\omega' = t - u\sqrt{d},$$

which is also in Ω. Thus Ω is a normal field (§16). To investigate the *integers* ω we put

$$t = \frac{x}{z}, \quad u = \frac{y}{z},$$

where x, y, z are rational integers with no common divisor and z can be assumed positive. Now if ω is an integer, so is ω' (§16) and consequently

$$\omega + \omega' = \frac{2x}{z}, \quad \omega\omega' = \frac{x^2 - dy^2}{z^2}$$

must also be integers. Conversely, if the latter are integers, then so is ω (§13). Suppose that e is the greatest common divisor of z and x. It is necessary that e^2 divide $x^2 - dy^2$, hence also dy^2 and finally y^2, since d is not divisible by any square other than 1. Then e must also divide y, and hence be equal to 1, since x, y, z have no common divisor. Thus z is prime to x while z divides $2x$, which means that $z = 1$ or $z = 2$. In the first case, $\omega = x + y\sqrt{d}$ is certainly an integer. In the second case x is odd, so $x^2 \equiv 1$ (mod 4), and since we must have $x^2 \equiv dy^2$ (mod 4) it is necessary that y also be odd, and therefore we have $d \equiv 1$ (mod 4). If the latter condition is not satisfied, that is, if $d \equiv 2$ or $d \equiv 3$ (mod 4), then z must equal 1 and so we have $\mathfrak{o} = [1, \sqrt{d}]$ and

$$D = \begin{vmatrix} 1 & \sqrt{d} \\ 1 & -\sqrt{d} \end{vmatrix}^2 = 4d.$$

But if we have $d \equiv 1$ (mod 4), z can also equal 2,† and we have

$$\mathfrak{o} = \left[1, \frac{1 + \sqrt{d}}{2} \right], \quad \text{and} \quad D = \begin{vmatrix} 1 & \frac{1+\sqrt{d}}{2} \\ 1 & \frac{1-\sqrt{d}}{2} \end{vmatrix}^2 = d.$$

These two cases can be combined into one by noting that $\mathfrak{o} = [1, \frac{D+\sqrt{D}}{2}]$ in each. It is also clear that a *quadratic* field is completely determined by its discriminant D. This is not so for the next case, namely $n = 3$, where invariants additional to the discriminant are necessary for the complete determination of a *cubic* field. However, we shall be able to give a complete explanation of this fact only with the help of the theory of *ideals*.

We now return to considering an arbitrary field Ω of degree n, and we add the following remarks on divisibility and congruence of numbers in the domain $\mathfrak{o}$. Let λ, μ be two such numbers, and suppose that λ is divisible by μ. By the general definition of divisibility (§14) we then have $\lambda = \mu\omega$, where ω is an integer, and since the quotient ω of the two numbers λ, μ belongs to Ω, by definition of a field, ω will likewise be a number in the domain $\mathfrak{o}$. The system $\mathfrak{m}$ of all numbers in the field Ω

---

† It follows, for example, in the case $d = -3$ that the integers of the field are not *all* of the form $x + y\sqrt{-3}$ where x, y are rational integers.

divisible by μ therefore consists of all numbers of the form $\mu\omega$, as ω runs through all numbers of the domain $\mathfrak{o} = [\omega_1, \omega_2, \ldots, \omega_n]$, that is, through all numbers of the form

$$\omega = h_1\omega_1 + h_2\omega_2 + \cdots + h_n\omega_n,$$

whose coordinates $h_1, h_2, \ldots, h_n$ are rational integers. Consequently we have $\mathfrak{m} = [\mu\omega_1, \mu\omega_2, \ldots, \mu\omega_n]$. We now say that two integers α, β in the domain $\mathfrak{o}$ are *congruent modulo* μ, and write

$$\alpha \equiv \beta \pmod{\mu},$$

when the difference $\alpha - \beta$ is divisible by μ and hence in $\mathfrak{m}$. Thus this congruence is completely equivalent to the following:

$$\alpha \equiv \beta \pmod{\mathfrak{m}},$$

the meaning of which was explained in §2. In the contrary case, α, β are called *incongruent* modulo μ. If we understand a *class* modulo μ to be the set of all numbers in $\mathfrak{o}$ congruent to a particular number, and hence congruent to each other, then, in the notation of §2, the number of these classes will be $(\mathfrak{o}, \mathfrak{m})$. And since the integers $\mu\omega_1, \mu\omega_2, \ldots, \mu\omega_n$ forming the basis of $\mathfrak{m}$ are connected to the numbers $\omega_1, \omega_2, \ldots, \omega_n$ by n equations of the form (6), (§17), in which the coefficients $m_{i,i'}$ are necessarily rational *integers*, it follows from the equation following (7), together with Theorem 4 of §4, that the number of these classes is

$$(\mathfrak{o}, \mathfrak{m}) = \pm N(\mu).$$

The system $\mathfrak{m}$ is identical with $\mathfrak{o}$ if and only if μ is a unit, in which case $\pm N(\mu) = (\mathfrak{o}, \mathfrak{o}) = 1$.

In this conception of congruence, where an actual number μ appears as divisor or modulus, there is a complete analogy with the theory of rational numbers. However, it is plain, as we have already indicated in detail in the Introduction and in Chapter 2, that completely new phenomena concerned with the decomposition of numbers into factors arise in this same domain $\mathfrak{o}$. These phenomena are brought back under the rule of simple laws by the *theory of ideals*, the elements of which will be covered in the next chapter.

4

Elements of the theory of ideals

In this chapter we shall develop the theory of ideals to the point indicated in the Introduction, that is, we shall prove the fundamental laws which apply to all fields of finite degree, and which regulate and explain the phenomena of divisibility in the domain o of all integers in such a field Ω. The latter are what we refer to when we speak of "numbers" in what follows, unless the contrary is expressly indicated. The theory is founded on the notion of *ideal*, whose origin has been mentioned in the Introduction, and whose importance has been sufficiently illuminated by the example in Chapter 2 (§§11 and 12). The exposition of the theory that follows coincides with the one I have given in the second edition of Dirichlet's *Vorlesungen über Zahlentheorie* (§163). It differs mainly in external form; however, if the theory has not been shortened it has at least been simplified a little. In particular, the principal difficulty to be surmounted is now thrown more clearly into relief.

§19. Ideals and their divisibility

As in the previous chapter, let Ω be a field of finite degree n, and let o be the domain of integers ω in Ω. An *ideal* of this domain o is a system $\mathfrak{a}$ of numbers α in o with the following two properties:

I. The sum and difference of any two numbers in $\mathfrak{a}$ also belong to $\mathfrak{a}$, that is, $\mathfrak{a}$ is a module.

II. The product $\alpha\omega$ of any number α in $\mathfrak{a}$ with a number ω in o is a number in $\mathfrak{a}$.

We begin by mentioning an important special case of the concept of ideal. Let μ be a particular number; then the system $\mathfrak{a}$ of all numbers $\alpha = \mu\omega$ divisible by μ forms an ideal. We call such an ideal a *principal*

119

ideal and denote it by $o(\mu)$, or more simply by $o\mu$ or μo. It is evident that this ideal will be unchanged when μ is replaced by an associate, that is, a number of the form $\epsilon\mu$, where ϵ is a unit. If μ is itself a unit we have $o\mu = o$, since all numbers in o are divisible by μ. It is easy to see that no other ideal can contain a unit. Because if the unit ϵ is in the ideal a then (by II) all products $\epsilon\omega$, and hence all numbers ω in the principal ideal o, are in a. But since, by definition, all numbers in the ideal a are likewise in o, we have $a = o$. The ideal o plays the same role among the ideals as the number 1 plays among the rational integers. The notion of principal ideal $o\mu$ also includes the singular case where $\mu = 0$, where the resulting ideal consists of the single number zero. However, we shall exclude this case from now on.

In the case $n = 1$, where our theory becomes the old theory of numbers, every ideal is evidently a principal ideal, that is, a module of the form $[m]$ where m is a rational integer (§§1 and 5). The same is true for the special quadratic fields considered in Chapter 2 (§6 and the beginning of §7). In all these cases, where every ideal of the field Ω is a principal ideal, numbers are governed by the same laws that govern the theory of rational integers, because every indecomposable number also has the character of a *prime number* (see the Introduction and §7). This will follow easily from the results below, but I mention it now to encourage the reader to make continual comparisons with the special cases, and especially with the old theory of rational numbers, because without doubt it will help greatly in understanding our general theory.

Since each ideal is a module (by virtue of I), we immediately carry over to ideals the notion of divisibility of modules (§1). We say that an ideal m is *divisible* by an ideal a, or that it is a *multiple* of a, when all the numbers in m are also in a. At the same time we say that a is a *divisor* of m. According to this definition, each ideal is divisible by the ideal o. If α is a number in the ideal a then the principal ideal $o\alpha$ will be divisible by a (by II). For this reason we say that the *number* α, and hence every number in a, is *divisible* by the ideal a.

Likewise we say that an ideal a is *divisible* by the *number* η when a is divisible by the principal ideal $o\eta$. When all the numbers α in an ideal a are of the form $\eta\rho$ it is easy to see that the system r of all the numbers $\rho = \alpha/\eta$ will form an ideal. Conversely, if ρ runs through all the numbers in an ideal r while η is a fixed nonzero number, then all the products $\eta\rho$ will again form an ideal, divisible by $o\eta$. We shall denote an ideal formed in this way from an ideal r and number η by $r\eta$ or ηr. We evidently have $(r\eta)\eta' = r(\eta\eta') = (r\eta')\eta$, and $\eta r'$ will be divisible by

$\eta \mathfrak{r}$ if and only if $\mathfrak{r}'$ is divisible by $\mathfrak{r}$. Thus the equation $\eta \mathfrak{r}' = \eta \mathfrak{r}$ entails the equation $\mathfrak{r}' = \mathfrak{r}$. The notion of a principal ideal $\mathfrak{o}\mu$ is the special case of $\mathfrak{r}\mu$ where $\mathfrak{r} = \mathfrak{o}$.

We finally remark that divisibility of the principal ideal $\mathfrak{o}\mu$ by the principal ideal $\mathfrak{o}\eta$ is completely equivalent to divisibility of the *number* μ by the *number* η. The laws of divisibility of *numbers* in $\mathfrak{o}$ are therefore included in the laws of divisibility of *ideals*.

The least common multiple $\mathfrak{m}$ and the greatest common divisor $\mathfrak{d}$ of two ideals $\mathfrak{a}$, $\mathfrak{b}$ are also ideals. Certainly $\mathfrak{m}$ and $\mathfrak{d}$ are modules (§1, 3 and 4), and they are divisible by $\mathfrak{o}$, since $\mathfrak{a}$ and $\mathfrak{b}$ are divisible by $\mathfrak{o}$. Also, if $\mu = \alpha = \beta$ is a number in $\mathfrak{m}$ and hence also in $\mathfrak{a}$ and $\mathfrak{b}$, and if $\delta = \alpha' + \beta'$ is a number in the module $\mathfrak{d}$ then the product $\mu\omega = \alpha\omega = \beta\omega$ will likewise be in $\mathfrak{m}$ and the product $\delta\omega = \alpha'\omega + \beta'\omega$ will be in $\mathfrak{d}$ since (by virtue of II) the products $\alpha\omega$, $\alpha'\omega$ are in $\mathfrak{a}$ and the products $\beta\omega$, $\beta'\omega$ are in $\mathfrak{b}$. Thus $\mathfrak{m}$ and $\mathfrak{d}$ enjoy all the properties of ideals. At the same time it is clear that $\mathfrak{m}\eta$ will be the least common multiple of the ideals $\mathfrak{a}\eta$, $\mathfrak{b}\eta$, and $\mathfrak{d}\eta$ their greatest common divisor.

If $\mathfrak{b}$ is a principal ideal $\mathfrak{o}\eta$, then the least common multiple $\mathfrak{m}$ of $\mathfrak{a}$, $\mathfrak{b}$ will always be of the form $\eta\mathfrak{r}$, where $\mathfrak{r}$ is another ideal and in fact a divisor of $\mathfrak{a}$, since $\eta\mathfrak{a}$ is a common multiple of $\mathfrak{a}$ and $\mathfrak{o}\eta$, and hence divisible by $\eta\mathfrak{r}$. This case occurs frequently in what follows, and for that reason we say, for brevity, that the ideal $\mathfrak{r}$ dividing the ideal $\mathfrak{a}$ *corresponds* to the number η. If $\mathfrak{r}'$ is the divisor of $\mathfrak{r}$ corresponding to the number η', then $\mathfrak{r}'$ will also be the divisor of $\mathfrak{a}$ corresponding to the product $\eta\eta'$. This is because $\eta\eta'\mathfrak{r}'$ is the least common multiple of $\eta\mathfrak{r}$ and $\mathfrak{o}\eta\eta'$, and hence also of $\mathfrak{a}$ and $\mathfrak{o}\eta\eta'$, since $\eta\mathfrak{r}$ is the least common multiple of $\mathfrak{a}$ and $\mathfrak{o}\eta$, and $\mathfrak{o}\eta\eta'$ is divisible by $\mathfrak{o}\eta$.

§20. Norms

Since each ideal $\mathfrak{a}$ is also a module, we say that two numbers ω, ω' in the domain $\mathfrak{o}$ are *congruent* or *incongruent modulo* $\mathfrak{a}$ according as their difference $\omega - \omega'$ belongs to $\mathfrak{a}$ or not. We express the congruence of ω and ω' modulo $\mathfrak{a}$ (§2) by the notation

$$\omega \equiv \omega' \pmod{\mathfrak{a}}.$$

As well as the theorems on congruences established previously for arbitrary modules, we must also note that two congruences modulo the same ideal $\mathfrak{a}$,

$$\omega \equiv \omega', \quad \omega'' \equiv \omega''' \quad (\text{mod } \mathfrak{a}),$$

can be multiplied to give the congruence

$$\omega\omega'' \equiv \omega'\omega''' \ (\text{mod } \mathfrak{a}),$$

since the products $(\omega - \omega')\omega''$ and $(\omega'' - \omega''')\omega'$, and hence also their sum $\omega\omega'' - \omega'\omega'''$, are numbers in the ideal $\mathfrak{a}$. Moreover, if $\mathfrak{m}$ is a principal ideal $\mathfrak{o}\mu$, then (by virtue of §18) the congruence $\omega \equiv \omega'$ (mod $\mathfrak{m}$) will be identical with the congruence $\omega \equiv \omega'$ (mod μ).

A very important consideration is the *number* of different classes, modulo the ideal $\mathfrak{a}$, which make up the domain $\mathfrak{o}$. If μ is a particular nonzero number in the ideal $\mathfrak{a}$ then the principal ideal $\mathfrak{o}\mu$ will be divisible by $\mathfrak{a}$, and since $\mathfrak{a}$ is divisible by $\mathfrak{o}$ it follows (§2,4) that

$$(\mathfrak{o}, \mathfrak{o}\mu) = (\mathfrak{o}, \mathfrak{a})(\mathfrak{a}, \mathfrak{o}\mu).$$

But the number $(\mathfrak{o}, \mathfrak{o}\mu) = \pm N(\mu)$ by §18, and hence the domain $\mathfrak{o}$ contains only a finite number of mutually incongruent numbers modulo the ideal $\mathfrak{a}$ (§2,2). This number $(\mathfrak{o}, \mathfrak{a})$ will be called the *norm of the ideal* $\mathfrak{a}$ and we denote it by $N(\mathfrak{a})$. The norm of the principal ideal $\mathfrak{o}\mu$ is equal to $\pm N(\mu)$, and $\mathfrak{o}$ is evidently the only ideal with norm 1.

If ρ runs through a complete system of $N(\mathfrak{a})$ incongruent numbers (mod $\mathfrak{a}$) then so does $(1 + \rho)$, and adding the corresponding congruences $1 + \rho \equiv \rho'$, where ρ' runs through the same values as ρ, yields $N(\mathfrak{a}) \equiv 0$ (mod $\mathfrak{a}$). That is, $N(\mathfrak{a})$ is always divisible by $\mathfrak{a}$. As a special case, this result includes the evident theorem that $N(\mu)$ is divisible by μ (see §17).

Now suppose that $\mathfrak{r}$ is any ideal and η is a nonzero number. Always,

$$(\mathfrak{o}\eta, \mathfrak{r}\eta) = (\mathfrak{o}, \mathfrak{r}) = N(\mathfrak{r}),$$

since two numbers $\eta\omega'$ and $\eta\omega''$ in the principal ideal $\eta\mathfrak{o}$ are congruent (mod $\eta\mathfrak{r}$) if and only if the numbers ω', ω'' in $\mathfrak{o}$ are congruent (mod $\mathfrak{r}$).

Let $\mathfrak{a}$, $\mathfrak{b}$ be any two ideals, let $\mathfrak{m}$ be their least common multiple, and let $\mathfrak{d}$ be their greatest common divisor. By §2, 3 and 4, we have

$$(\mathfrak{b}, \mathfrak{a}) = (\mathfrak{b}, \mathfrak{m}) = (\mathfrak{d}, \mathfrak{a})$$

and, since $\mathfrak{d}$ is divisible by $\mathfrak{o}$,

$$(\mathfrak{o}, \mathfrak{a}) = (\mathfrak{o}, \mathfrak{d})(\mathfrak{d}, \mathfrak{a}), \quad (\mathfrak{o}, \mathfrak{m}) = (\mathfrak{o}, \mathfrak{b})(\mathfrak{b}, \mathfrak{m}),$$

hence

$$N(\mathfrak{a}) = (\mathfrak{b}, \mathfrak{a})N(\mathfrak{d}), \quad N(\mathfrak{m}) = (\mathfrak{b}, \mathfrak{a})N(\mathfrak{b})$$

and

$$N(\mathfrak{m})N(\mathfrak{d}) = N(\mathfrak{a})N(\mathfrak{b}).$$

If we apply these theorems to the case where $\mathfrak{b}$ is a principal ideal $\mathfrak{o}\eta$, so that $\mathfrak{m}$ is of the form $\mathfrak{r}\eta$, then, since the ideal $\mathfrak{r}$ is the divisor of $\mathfrak{a}$ corresponding to the number η (§19), we get

$$(\mathfrak{b}, \mathfrak{a}) = (\mathfrak{o}\eta, \mathfrak{r}\eta) = N(\mathfrak{r}),$$

and consequently

$$N(\mathfrak{a}) = N(\mathfrak{r})N(\mathfrak{d}).$$

The ideal $\mathfrak{r}$ can also be defined as the system of all roots ρ of the congruence $\eta\rho \equiv 0 \pmod{\mathfrak{a}}$, as is easy to see.

§*21. Prime ideals*

An ideal $\mathfrak{p}$ is called *prime* when it is different from $\mathfrak{o}$ and divisible by no ideals except $\mathfrak{o}$ and $\mathfrak{p}$. This definition yields the following theorems:

1. Each ideal $\mathfrak{a}$ different from $\mathfrak{o}$ is divisible by a prime ideal.

This is because the ideals different from $\mathfrak{o}$ that divide the ideal $\mathfrak{a}$ include one, $\mathfrak{p}$, whose norm is *smallest*, and the latter is certainly a prime ideal. Indeed, if $\mathfrak{d}$ is an ideal dividing $\mathfrak{p}$, but different from $\mathfrak{p}$ and $\mathfrak{o}$, then we have $(\mathfrak{d}, \mathfrak{p}) > 1$. Hence $N(\mathfrak{p}) = (\mathfrak{d}, \mathfrak{p})N(\mathfrak{d}) > N(\mathfrak{d})$ and $\mathfrak{d}$ will be a divisor of the ideal $\mathfrak{a}$, different from $\mathfrak{o}$ and with norm $< N(\mathfrak{p})$, contrary to hypothesis. Thus $\mathfrak{p}$ is a prime ideal.   Q.E.D.

2. If the number η is not divisible by the prime ideal $\mathfrak{p}$, then $\eta\mathfrak{p}$ will be the least common multiple of the two ideals $\mathfrak{p}$ and $\mathfrak{o}\eta$.

The least common multiple of $\mathfrak{p}$ and $\mathfrak{o}\eta$ is in any case of the form $\eta\mathfrak{r}$, with the ideal $\mathfrak{r}$ a divisor of $\mathfrak{a}p$ and hence equal to $\mathfrak{o}$ or $\mathfrak{p}$. But $\mathfrak{r}$ cannot be $\mathfrak{o}$, because $\eta\mathfrak{o}$ is not divisible by $\mathfrak{p}$; consequently $\mathfrak{r} = \mathfrak{p}$.   Q.E.D.

3. If neither of the two numbers η, ρ is divisible by the prime ideal $\mathfrak{p}$, then their product $\eta\rho$ will not be divisible by $\mathfrak{p}$.

Otherwise the ideal $\eta(\mathfrak{o}\rho)$ would be a common multiple of $\mathfrak{p}$ and $\mathfrak{o}\eta$. Consequently it would be divisible by the least common multiple $\eta\mathfrak{p}$ of $\mathfrak{p}$ and $\mathfrak{o}\eta$. But the divisibility of $\eta(\mathfrak{o}\rho)$ by $\eta\mathfrak{p}$ implies (§19) the divisibility of $\mathfrak{o}\rho$ by $\mathfrak{p}$, contrary to hypothesis. Thus $\eta\rho$ is not divisible by $\mathfrak{p}$. Q.E.D.

It follows immediately from this that all the *rational* numbers divisible by a prime ideal $\mathfrak{p}$, amongst which we have the number $N(\mathfrak{p})$ (§20), form a module $[p]$, where p is a positive rational *prime number*. This is because the *smallest* positive rational p divisible by $\mathfrak{p}$ cannot be a composite number ab, otherwise one of the smaller numbers a, b would likewise be divisible by $\mathfrak{p}$, and p cannot be 1 otherwise we should have $\mathfrak{p} = \mathfrak{o}$ (§19). And each rational integer m divisible by $\mathfrak{p}$ must be divisible by p, as becomes evident by putting m in the form $pq + r$, since the remainder $r = m - pq$ is also divisible by $\mathfrak{p}$. Now with $\mathfrak{o}p$ being divisible by $\mathfrak{p}$, and hence $N(\mathfrak{o}p) = p^n$ being divisible by $N(\mathfrak{p})$ (§20), $N(\mathfrak{p})$ will be a power p^f of p, and the exponent f will be called the *degree of the prime ideal* $\mathfrak{p}$.

4. If the ideal $\mathfrak{a}$ is divisible by the prime ideal $\mathfrak{p}$ then there is a number η such that $\eta\mathfrak{p}$ is the least common multiple of $\mathfrak{a}$ and $\mathfrak{o}\eta$.

This important theorem is evident when we have $\mathfrak{a} = \mathfrak{p}$, because any number η not divisible by $\mathfrak{p}$, for example $\eta = 1$, satisfies the condition. But if $\mathfrak{a}$ is different from $\mathfrak{p}$ we first confine ourselves to showing the existence of a number η such that the divisor $\mathfrak{r}$ of the ideal $\mathfrak{a}$ corresponding to η is at the same time divisible by $\mathfrak{p}$, while having norm *less* than that of $\mathfrak{a}$. Since we have $N(\mathfrak{a}) = N(\mathfrak{r})N(\mathfrak{d})$, where $\mathfrak{d}$ is the greatest common divisor of $\mathfrak{a}$ and $\mathfrak{o}\eta$ (§20), the latter condition depends on choosing η so that $N(\mathfrak{d}) > 1$, which means that $\mathfrak{d}$ is different from $\mathfrak{o}$. To attain this goal, and at the same time make $\mathfrak{r}$ divisible by $\mathfrak{p}$, we distinguish two cases:

First, if all the ideals (except $\mathfrak{o}$) that divide $\mathfrak{a}$ are divisible by $\mathfrak{p}$, then we choose η to be a number divisible by $\mathfrak{p}$ but not divisible by $\mathfrak{a}$, which is always possible because $\mathfrak{p}$ is not divisible by $\mathfrak{a}$. Then it is clear that $\mathfrak{d}$ will be divisible by $\mathfrak{p}$, and hence will be different from $\mathfrak{o}$. Moreover, since η is not divisible by $\mathfrak{a}$, and $\eta\mathfrak{r}$ is, $\mathfrak{r}$ will also be different from $\mathfrak{o}$, and hence divisible by $\mathfrak{p}$.

Second, if there is an ideal $\mathfrak{e}$ different from $\mathfrak{o}$ which divides $\mathfrak{a}$ and is not divisible by $\mathfrak{p}$, then we choose η to be a number divisible by $\mathfrak{e}$ but not divisible by $\mathfrak{p}$. Then $\mathfrak{d}$ will be divisible by $\mathfrak{e}$ and hence again different from $\mathfrak{o}$. Moreover, since $\eta\mathfrak{r}$ is divisible by $\mathfrak{a}$ and hence also by $\mathfrak{p}$, $\mathfrak{r}$ will also be divisible by $\mathfrak{p}$, because η is not divisible by $\mathfrak{p}$ (by 1).

Having established the existence in both cases of at least one number η with the required property, we see without difficulty that we certainly have $\mathfrak{r} = \mathfrak{p}$ if we choose η so that $N(\mathfrak{r})$ is *as small as possible*. Because, if the ideal $\mathfrak{r}$ divisible by $\mathfrak{p}$ is not equal to $\mathfrak{p}$, we can proceed with $\mathfrak{r}$ the same way as with $\mathfrak{a}$, and choose a number η' in such a way that the divisor $\mathfrak{r}'$ of $\mathfrak{r}$ corresponding to η' again has norm less than that of $\mathfrak{r}$, while likewise being divisible by $\mathfrak{p}$. But since (§19) $\mathfrak{r}'$ is at the same time the divisor of $\mathfrak{a}$ corresponding to the number $\eta\eta'$ this contradicts the assumption we have made about η and $\mathfrak{r}$. Thus $\mathfrak{r} = \mathfrak{p}$, that is, $\eta\mathfrak{p}$ is the least common multiple of $\mathfrak{a}$ and $\mathfrak{o}\eta$.   Q.E.D.

§22. Multiplication of ideals

If α runs through all the numbers in an ideal $\mathfrak{a}$, and β runs through those of an ideal $\mathfrak{b}$, then all the products of the form $\alpha\beta$, together with their sums, form an ideal $\mathfrak{c}$. These numbers are in $\mathfrak{o}$ and they are closed under addition; also under subtraction, because the numbers $(-\alpha)$ are likewise in $\mathfrak{a}$. Finally, each product of a number $\Sigma\alpha\beta$ in $\mathfrak{c}$ by a number ω in $\mathfrak{o}$ also belongs to $\mathfrak{c}$, since each product $\alpha\omega$ again belongs to $\mathfrak{a}$. This ideal $\mathfrak{c}$ is called the *product* of the two *factors* $\mathfrak{a}$, $\mathfrak{b}$, and we denote it by $\mathfrak{ab}$.

It follows immediately from this definition that $\mathfrak{oa} = \mathfrak{a}$, $\mathfrak{ab} = \mathfrak{ba}$ and, if $\mathfrak{c}$ is any third ideal, $(\mathfrak{ab})\mathfrak{c} = \mathfrak{a}(\mathfrak{bc})$, whence we conclude by well-known arguments† that in a product of any number of ideals $\mathfrak{a}_1, \mathfrak{a}_2, \ldots, \mathfrak{a}_m$ the order of the multiplications, which combine *two* ideals into a single product, has no influence on the final result, which can be written simply as $\mathfrak{a}_1\mathfrak{a}_2 \cdots \mathfrak{a}_m$ and evidently consists of all numbers of the form $\Sigma\alpha_1\alpha_2 \ldots \alpha_m$, where $\alpha_1, \alpha_2, \ldots, \alpha_m$ are numbers from the respective factors $\mathfrak{a}_1, \mathfrak{a}_2, \ldots, \mathfrak{a}_m$. If all the m factors equal $\mathfrak{a}$ their product will be called the mth *power* of $\mathfrak{a}$, and we denote it by $\mathfrak{a}^m$. We also put $\mathfrak{a}^0 = \mathfrak{o}$, $\mathfrak{a}^1 = \mathfrak{a}$ and in general we have $\mathfrak{a}^r\mathfrak{a}^s = \mathfrak{a}^{r+s}$, $(\mathfrak{a}^r)^s = \mathfrak{a}^{rs}$. In addition, we evidently have $\mathfrak{a}(\mathfrak{o}\eta) = \mathfrak{a}\eta$ and $(\mathfrak{o}\eta)(\mathfrak{o}\eta') = \mathfrak{o}\eta\eta'$. Finally, we re-establish the following theorems:

1. The product $\mathfrak{ab}$ is divisible by the factors $\mathfrak{a}$ and $\mathfrak{b}$.

This is because (by virtue of property II) each product $\alpha\omega$, hence each product $\alpha\beta$, and hence (by I) each sum of such products belongs to $\mathfrak{a}$. That is, $\mathfrak{ab}$ is divisible by $\mathfrak{a}$.

2. If $\mathfrak{a}$ is divisible by $\mathfrak{a}'$, and $\mathfrak{b}$ by $\mathfrak{b}'$, then $\mathfrak{ab}$ is divisible by $\mathfrak{a}'\mathfrak{b}'$.

This is because all the numbers $\Sigma\alpha\beta$ in $\mathfrak{ab}$ are in $\mathfrak{a}'\mathfrak{b}'$, since α is in $\mathfrak{a}$ and hence in $\mathfrak{a}'$, and β is in $\mathfrak{b}$ and hence in $\mathfrak{b}'$.

† See §2 of Dirichlet's *Vorlesungen über Zahlentheorie*.

3. If neither of the ideals $\mathfrak{a}$, $\mathfrak{b}$ is divisible by the prime ideal $\mathfrak{p}$, then the product $\mathfrak{ab}$ is also not divisible by $\mathfrak{p}$.

This is because there are numbers α, β, in $\mathfrak{a}$, $\mathfrak{b}$ respectively, which are not divisible by $\mathfrak{p}$, and then the number $\alpha\beta$ in $\mathfrak{ab}$ is also not divisible by $\mathfrak{p}$ (§21,3).

§23. The difficulty in the theory

It would be easy to augment considerably the number of theorems connecting the notions of *divisibility* and *multiplication* of ideals, and we mention without proof the following propositions, simply to emphasize the resemblance with the corresponding propositions in the theory of rational numbers.

If $\mathfrak{a}$, $\mathfrak{b}$ are *relatively prime* ideals, that is, having greatest common divisor $\mathfrak{o}$, then their least common multiple is $\mathfrak{ab}$, and at the same time

$$N(\mathfrak{ab}) = N(\mathfrak{a})N(\mathfrak{b}).$$

If $\mathfrak{p}$ is a prime ideal, and $\mathfrak{a}$ is any ideal, then either $\mathfrak{a}$ is divisible by $\mathfrak{p}$ or $\mathfrak{a}$ and $\mathfrak{p}$ are relatively prime.

If $\mathfrak{a}$ is an ideal relatively prime to $\mathfrak{b}$ and $\mathfrak{c}$, then $\mathfrak{a}$ is also relatively prime to $\mathfrak{bc}$.

If $\mathfrak{ab}$ is divisible by $\mathfrak{c}$ and $\mathfrak{a}$ is relatively prime to $\mathfrak{c}$, then $\mathfrak{b}$ is divisible by $\mathfrak{c}$.

However, all these propositions are insufficient to complete the analogy with the theory of rational numbers. It is necessary to remember that divisibility of an ideal $\mathfrak{c}$ by an ideal $\mathfrak{a}$, according to our definition (§19), means only that all the numbers in the ideal $\mathfrak{c}$ are also in the ideal $\mathfrak{a}$. It is very easy to see (§22,1) that any product of $\mathfrak{a}$ by an ideal $\mathfrak{b}$ is divisible by $\mathfrak{a}$, but it is by no means easy to show the converse, that each ideal divisible by $\mathfrak{a}$ is the product of $\mathfrak{a}$ by an ideal $\mathfrak{b}$. This difficulty, which is the greatest and really the only one presented by the theory, cannot be surmounted by the methods we have employed thus far, and it is necessary to examine more closely the reason for this phenomenon, because it is connected with a very important generalisation of the theory. By attentively considering the theory developed until now, one notices that all the definitions retain their meaning, and the proofs of all theorems still hold, when one *no longer supposes* that the domain $\mathfrak{o}$ consists of *all* integers in the field Ω. The only properties of $\mathfrak{o}$ really needed are the following:

(a) The system $\mathfrak{o}$ is a finitely generated module $[\omega_1, \omega_2, \ldots, \omega_n]$ whose basis is also a basis for the field Ω.

(b) The number 1 is in $\mathfrak{o}$, hence so are all the rational integers.

(c) Each product of two numbers in $\mathfrak{o}$ is also in $\mathfrak{o}$.

When a domain $\mathfrak{o}$ enjoys these three properties we shall call it an *order*. It follows immediately from (a) and (c) that an order consists entirely of *integers* from the field Ω, but it does not necessarily contain *all* the integers (except in the case $n = 1$). Now if a number α in the order $\mathfrak{o}$ is called *divisible* by a second such number μ only when $\alpha = \mu\omega$, with ω also in $\mathfrak{o}$, and if we modify the notion of *congruence* of numbers in $\mathfrak{o}$ in the same manner, then one sees immediately that the number $(\mathfrak{o}, \mathfrak{o}\mu)$ of mutually incongruent numbers of $\mathfrak{o}$ modulo μ is again $\pm N(\mu)$ (§18). It is also easy to see that all the definitions and all the theorems of the present chapter retain their meaning and truth if we always understand *number* to mean a number in the order $\mathfrak{o}$. Thus, in particular, each order $\mathfrak{o}$ in field Ω has its own theory of ideals, and this theory is the same for all orders (which are infinite in number) up to the point we have carried it so far. However, while the theory of ideals in the order $\mathfrak{o}$ of *all* integers in the field Ω leads finally to general laws which coincide completely with the laws of divisibility for the rational numbers, the theory of ideals in other orders is subject to certain exceptions, or rather, it requires a certain limitation of the notion of ideal. The general theory of ideals in an arbitrary order, whose development is equally indispensable for the theory of numbers and which, in the case $n = 2$, coincides with the theory of *orders* of binary quadratic *forms*,† will be left aside in what follows,‡ and I shall content myself with giving an example to call attention to the character of the exceptions just mentioned. In the quadratic field resulting from a root

$$\theta = \frac{-1 + \sqrt{-3}}{2}$$

of the equation $\theta^2 + \theta + 1 = 0$ the module $[1, \sqrt{-3}]$ forms an order $\mathfrak{o}$ not containing all the integers of the field. The modules $[2, 1+\sqrt{-3}] = \mathfrak{p}$ and $[2, 2\sqrt{-3}] = \mathfrak{o}(2)$ must be regarded as ideals in this order $\mathfrak{o}$, inasmuch as they enjoy the properties I and II (§19). However, while $\mathfrak{o}(2)$ is divisible by $\mathfrak{p}$, there is no ideal $\mathfrak{q}$ in $\mathfrak{o}$ such that $\mathfrak{p}\mathfrak{q} = \mathfrak{o}(2)$.

† *Disquisitiones Arithmeticae*, art. 226.

‡ I treat this theory in detail in the recently published memoir: "*Ueber die Anzahl der Ideal-Classen in den verschiedenen Ordnungen eines endlichen Körpers.*" (*Festschrift zur Säcularfeier des Geburtstages von C.-F. Gauss.* Braunschweig, 30 April 1877).

§24. *Auxiliary propositions*

To achieve the completion of the theory of ideals in an order o containing *all* the integers of a field Ω we need the following lemmas, which are not true unless we restrict ourselves to such a domain o.

1. Let ω, μ, ν be three nonzero numbers in o such that ν is not divisible by μ. Then the terms of the geometric progression

$$\omega, \quad \omega\frac{\nu}{\mu}, \quad \omega\left(\frac{\nu}{\mu}\right)^2, \quad \omega\left(\frac{\nu}{\mu}\right)^3, \quad \ldots,$$

up to a term

$$\omega\left(\frac{\nu}{\mu}\right)^e$$

at some finite position, will all be in o, and beyond that none of them will be an integer.

In fact, if the number of integral terms exceeds the absolute value k of $N(\omega)$ then there must be (by §18) at least two among them, corresponding to exponents s and $r > s$, which are congruent modulo ω. But such a congruence,

$$\omega\left(\frac{\nu}{\mu}\right)^r \equiv \omega\left(\frac{\nu}{\mu}\right)^s \pmod{\omega},$$

implies that the number

$$\eta = \frac{\nu}{\mu}$$

in the field Ω satisfies an rth degree equation of the form

$$\eta^r = \eta^s + \omega'$$

where ω' is an integer, and hence (§13,2) η itself is an integer, contrary to our hypothesis that ν is not divisible by μ. Thus at most k terms of the series can be integers, and hence members of o. Moreover, if the term

$$\rho = \omega\left(\frac{\nu}{\mu}\right)^r,$$

with $r \geq 1$, is an integer, and if s is any one of the r exponents $0, 1, 2, \ldots, r-1$ then the term

$$\sigma = \omega\left(\frac{\nu}{\mu}\right)^s$$

will also be an integer, because

$$\sigma^r = \omega^{r-s}\rho^s$$

is an integer (§13,2). This completes the proof of the proposition.

2. Let μ, ν be two nonzero numbers in $\mathfrak{o}$, with ν not divisible by μ. Then there are two nonzero numbers κ, λ in $\mathfrak{o}$ such that

$$\frac{\kappa}{\lambda} = \frac{\nu}{\mu}$$

and κ^2 is not divisible by λ.

If

$$\lambda = \mu \left(\frac{\nu}{\mu}\right)^{e-1}, \quad \kappa = \mu \left(\frac{\nu}{\mu}\right)^e$$

are the last two integral terms of the series

$$\mu, \quad \mu\left(\frac{\nu}{\mu}\right), \quad \mu\left(\frac{\nu}{\mu}\right)^2, \quad \mu\left(\frac{\nu}{\mu}\right)^3, \quad \ldots,$$

and hence in $\mathfrak{o}$, then we evidently have $e \geq 1$ and

$$\frac{\kappa}{\lambda} = \frac{\nu}{\mu}, \quad \frac{\kappa^2}{\lambda} = \mu\left(\frac{\nu}{\mu}\right)^{e+1}.$$

Thus κ^2 is not divisible by λ. Q.E.D.

§25. Laws of divisibility

With the aid of these lemmas it is easy to bring the theory of ideals in the domain $\mathfrak{o}$ to the desired conclusion, which is found in the following laws:

1. If $\mathfrak{p}$ is a prime ideal then there is a number λ divisible by $\mathfrak{p}$, and a number κ not divisible by $\mathfrak{p}$, such that $\kappa\mathfrak{p}$ is the least common multiple of $\mathfrak{o}\lambda$ and $\mathfrak{o}\kappa$.

Proof. Let μ be any nonzero number in the prime ideal $\mathfrak{p}$. Since $\mathfrak{o}\mu$ is divisible by $\mathfrak{p}$ there is a number ν such that $\nu\mathfrak{p}$ is the least common multiple of $\mathfrak{o}\mu$ and $\mathfrak{o}\nu$ (§21,4). This number ν cannot be divisible by μ, otherwise the least common multiple of $\mathfrak{o}\mu$ and $\mathfrak{o}\nu$ would be $\mathfrak{o}\nu$ and not $\mathfrak{p}\nu$. Now if we choose (§24,2) the two numbers κ, λ so that $\kappa\mu = \lambda\nu$ and κ^2 is not divisible by λ, then (§19) the ideal $\kappa\nu\mathfrak{p}$ will be the least common multiple of $\kappa(o\mu) = o\lambda\nu$ and $o\kappa\nu$. It follows (§19) that $\kappa\mathfrak{p}$ is the least common multiple of $o\lambda$ and $o\kappa$. Then $\mathfrak{p}$ is the divisor of the principal ideal $o\lambda$ corresponding to the number κ. However, κ is not divisible by $\mathfrak{p}$, otherwise κ^2 would be divisible by $\kappa\mathfrak{p}$ and hence also by λ.

2. Each prime ideal $\mathfrak{p}$ has a multiple, by an ideal $\mathfrak{d}$, which is a principal ideal.

Proof. We retain κ and λ from above with the same meaning, and let $\mathfrak{d}$ be the greatest common divisor of $o\lambda$ and $o\kappa$. We shall show that $\mathfrak{p}\mathfrak{d} = o\lambda$. In fact, since all numbers in the ideal $\mathfrak{d}$ are of the form $\delta = \kappa\omega + \lambda\omega'$, where ω, ω' are two numbers from o, if ϖ is any number in $\mathfrak{p}$ we have $\varpi\delta = \kappa\varpi\omega + \lambda\varpi\omega' \equiv 0 \pmod{\lambda}$, because $\kappa\mathfrak{p}$ and hence also $\kappa\varpi$ is divisible by $o\lambda$. Thus $\mathfrak{p}\mathfrak{d}$ is divisible by $o\lambda$. Conversely, if κ is not divisible by $\mathfrak{p}$, in which case o is the greatest common divisor of $o\kappa$ and $\mathfrak{p}$, we can express the number 1 in o as $\kappa\omega + \varpi$, with ω in $\mathfrak{p}$ and ϖ in $\mathfrak{p}$. We then have $\lambda = \lambda \cdot \kappa\omega + \varpi \cdot \lambda \equiv 0 \pmod{\mathfrak{p}\mathfrak{d}}$, because the first factors λ, ϖ are in $\mathfrak{p}$ and the second factors $\kappa\omega, \lambda$ are in $\mathfrak{d}$. Thus each of the two ideals $\mathfrak{p}\mathfrak{d}$, $o\lambda$ is divisible by the other, and consequently $\mathfrak{p}\mathfrak{d} = o\lambda$.   Q.E.D.

3. If the ideal $\mathfrak{a}$ is divisible by the prime ideal $\mathfrak{p}$ then there is exactly one ideal $\mathfrak{a}'$ such that $\mathfrak{p}\mathfrak{a}' = \mathfrak{a}$, and at the same time $N(\mathfrak{a}') < N(\mathfrak{a})$.

Proof. Suppose, as we now can, that $\mathfrak{p}\mathfrak{d} = o\lambda$. Then if $\mathfrak{a}$ is divisible by $\mathfrak{p}$, and hence $\mathfrak{a}\mathfrak{d}$ by $\mathfrak{p}\mathfrak{d}$ (§22,2), we have $\mathfrak{a}\mathfrak{d} = \lambda\mathfrak{a}'$ where $\mathfrak{a}'$ is an ideal (§19). Multiplying by $\mathfrak{p}$, we get $\lambda\mathfrak{a} = \lambda\mathfrak{p}\mathfrak{a}'$ and hence also $\mathfrak{a} = \mathfrak{p}\mathfrak{a}'$. Now let $\mathfrak{b}$ be an ideal also satisfying the condition $\mathfrak{p}\mathfrak{b} = \mathfrak{a}$. Multiplying the equation $\mathfrak{p}\mathfrak{b} = \mathfrak{p}\mathfrak{a}'$ by $\mathfrak{d}$ we get $\lambda\mathfrak{b} = \lambda\mathfrak{a}'$, whence $\mathfrak{b} = \mathfrak{a}'$. In addition, there is (§21,4) a number η such that $\eta\mathfrak{p}$ is the least common multiple of $\mathfrak{a}$ and $o\eta$. But with $\eta\mathfrak{p}$ being divisible by $\mathfrak{a} = \mathfrak{a}'\mathfrak{p}$ it follows, multiplying by $\mathfrak{d}$, that $o\eta\lambda$ is divisible by $\lambda\mathfrak{a}'$, and hence that η is divisible by $\mathfrak{a}'$. However, η is certainly not divisible by $\mathfrak{a}$, otherwise $o\eta$ and not $\eta\mathfrak{p}$ would be the least common multiple of $\mathfrak{a}$ and $o\eta$. Thus, since η is divisible by $\mathfrak{a}'$ and not by $\mathfrak{a}$, it follows that $\mathfrak{a}'$ is *different* from $\mathfrak{a}$, and consequently we have $N(\mathfrak{a}') < N(\mathfrak{a})$, because $\mathfrak{a}'$ is a divisor of $\mathfrak{a}$.   Q.E.D.

4. Each ideal $\mathfrak{a}$ different from o is either a prime ideal or else expressible as a product of prime ideals, and in only one way.

Proof. Since $\mathfrak{a}$ is different from o there is (§21,1) a prime ideal $\mathfrak{p}_1$ dividing $\mathfrak{a}$, and hence we can put (by 3) $\mathfrak{a} = \mathfrak{p}_1\mathfrak{a}_1$, where $N(\mathfrak{a}_1) < N(\mathfrak{a})$.

If we have $\mathfrak{a}_1 = \mathfrak{o}$ then $\mathfrak{a} = \mathfrak{p}_1$ will be a prime ideal. But if $N(\mathfrak{a}_1)$ is > 1, so that $\mathfrak{a}_1$ is different from $\mathfrak{o}$, we can likewise put $\mathfrak{a}_1 = \mathfrak{p}_2\mathfrak{a}_2$ where $\mathfrak{p}_2$ is a prime ideal and $N(\mathfrak{a}_2) < N(\mathfrak{a}_1)$. If $N(\mathfrak{a}_2) > 1$ we can continue in the same manner, obtaining ideals $\mathfrak{a}_1, \mathfrak{a}_2, \mathfrak{a}_3, \ldots$, whose norms are smaller and smaller, finally arriving at the ideal $\mathfrak{o} = \mathfrak{a}_m$ after a finite number of decompositions. We then have a product of m prime ideals

$$\mathfrak{a} = \mathfrak{p}_1\mathfrak{p}_2 \cdots \mathfrak{p}_m.$$

Now if at the same time

$$\mathfrak{a} = \mathfrak{q}_1\mathfrak{q}_2 \cdots \mathfrak{q}_m,$$

where $\mathfrak{q}_1, \mathfrak{q}_2, \ldots, \mathfrak{q}_m$ are also prime ideals, then $\mathfrak{q}_1$ will be a divisor of the product $\mathfrak{p}_1\mathfrak{p}_2 \cdots \mathfrak{p}_m$ and hence (§22,3) at least one of its factors, $\mathfrak{p}_1$ for example, must be divisible by $\mathfrak{q}_1$. And since $\mathfrak{p}_1$ is divisible only by the two ideals $\mathfrak{o}$ and $\mathfrak{p}_1$, we necessarily have $\mathfrak{q}_1 = \mathfrak{p}_1$, since $\mathfrak{q}_1$ is different from $\mathfrak{o}$. We then have

$$\mathfrak{p}_1(\mathfrak{p}_2\mathfrak{p}_3 \cdots \mathfrak{p}_m) = \mathfrak{p}_1(\mathfrak{q}_2\mathfrak{q}_3 \cdots \mathfrak{q}_m),$$

whence (by 3)

$$\mathfrak{p}_2\mathfrak{p}_3 \cdots \mathfrak{p}_m = \mathfrak{q}_2\mathfrak{q}_3 \cdots \mathfrak{q}_m,$$

We can continue in the same manner, just as in the theory of rational numbers,† and thus arrive at the result that each prime ideal factor in one product occurs in the other, and exactly the same number of times. Q.E.D.

5. Each ideal $\mathfrak{a}$, when multiplied by a suitable ideal $\mathfrak{m}$, becomes a principal ideal.

Proof. Suppose that $\mathfrak{a} = \mathfrak{p}_1\mathfrak{p}_2 \cdots \mathfrak{p}_m$. By 2, we can multiply the prime ideals $\mathfrak{p}_1, \mathfrak{p}_2, \ldots, \mathfrak{p}_m$ by the corresponding ideals $\mathfrak{d}_1, \mathfrak{d}_2, \ldots, \mathfrak{d}_m$ to change them into principal ideals $\mathfrak{p}_1\mathfrak{d}_1, \mathfrak{p}_2\mathfrak{d}_2, \ldots, \mathfrak{p}_m\mathfrak{d}_m$. If we now put

$$\mathfrak{m} = \mathfrak{d}_1\mathfrak{d}_2 \cdots \mathfrak{d}_m,$$

then $\mathfrak{am} = (\mathfrak{p}_1\mathfrak{d}_1)(\mathfrak{p}_2\mathfrak{d}_2) \cdots (\mathfrak{p}_m\mathfrak{d}_m)$ will be a unique product of principal ideals, and hence itself a principal ideal. Q.E.D.

6. If the ideal $\mathfrak{c}$ is divisible by the ideal $\mathfrak{a}$ then there is an ideal $\mathfrak{b}$, and only one, satisfying the condition $\mathfrak{ab} = \mathfrak{c}$. If the product $\mathfrak{ab}$ is divisible by the product $\mathfrak{ab}'$ then $\mathfrak{b}$ will be divisible by $\mathfrak{b}'$, and $\mathfrak{ab} = \mathfrak{ab}'$ implies $\mathfrak{b} = \mathfrak{b}'$.

† See Dirichlet's *Vorlesungen über Zahlentheorie*, §8.

Proof. We choose the ideal $\mathfrak{m}$ so that $\mathfrak{am}$ is a principal ideal $\mathfrak{o}\mu$. Now if $\mathfrak{c}$ is divisible by $\mathfrak{a}$, and hence $\mathfrak{cm}$ by $\mathfrak{am}$ (§22,2), we can, by §19, put $\mathfrak{cm} = \mu\mathfrak{b}$, where $\mathfrak{b}$ is an ideal. Multiplying by $\mathfrak{a}$, we get $\mu\mathfrak{c} = \mu\mathfrak{ab}$, whence $\mathfrak{c} = \mathfrak{ab}$. Now let $\mathfrak{a}$, $\mathfrak{b}$, $\mathfrak{b}'$ be any ideals, and suppose that $\mathfrak{ab}$ is divisible by $\mathfrak{ab}'$. Again it follows, multiplying by $\mathfrak{m}$ (§22,2), that $\mu\mathfrak{b}$ is divisible by $\mu\mathfrak{b}'$, and hence (§19) that $\mathfrak{b}$ is divisible by $\mathfrak{b}'$. If in addition $\mathfrak{ab} = \mathfrak{ab}'$, then each of the ideals $\mathfrak{b}$, $\mathfrak{b}'$ will be divisible by the other, that is, $\mathfrak{b} = \mathfrak{b}'$. Q.E.D.

7. The norm of a product of ideals is equal to the product of the norms of the factors: $N(\mathfrak{ab}) = N(\mathfrak{a})N(\mathfrak{b})$.

Proof. We first consider the case of a product $\mathfrak{a} = \mathfrak{pa}'$, where the factor $\mathfrak{p}$ is a prime ideal. Since $\mathfrak{a}$ is divisible by $\mathfrak{p}$, there is (by 3) a number η divisible by $\mathfrak{a}'$ but not by $\mathfrak{a}$, and $\eta\mathfrak{p}$ is the least common multiple of $\mathfrak{a}$ and $\mathfrak{o}\eta$. Then we have (§20) $N(\mathfrak{a}) = N(\mathfrak{p})N(\mathfrak{d})$, where $\mathfrak{d}$ is the greatest common divisor of the ideals $\mathfrak{a}$ and $\mathfrak{o}\eta$. Since $\mathfrak{a}$ and $\mathfrak{o}\eta$ are divisible by $\mathfrak{a}'$, $\mathfrak{d}$ must also be divisible by $\mathfrak{a}'$ (§1,4), and hence there is (by 6) an ideal $\mathfrak{n}$ satisfying the condition $\mathfrak{na}' = \mathfrak{d}$. Moreover, since $\mathfrak{a}$ is divisible by $\mathfrak{d}$, $\mathfrak{pa}'$ is divisible by $\mathfrak{na}'$, so the prime ideal $\mathfrak{p}$ must (by 6) be divisible by $\mathfrak{n}$, and thus we must have $\mathfrak{n} = \mathfrak{p}$ or $\mathfrak{o}$. The first equation is impossible unless we have $\mathfrak{d} = \mathfrak{pa}' = \mathfrak{a}$, so that η is divisible by $\mathfrak{a}$, and this is not the case. We therefore have $\mathfrak{n} = \mathfrak{o}$, whence $\mathfrak{d} = \mathfrak{a}'$ and also $N(\mathfrak{pa}') = N(\mathfrak{p})N(\mathfrak{a}')$, which proves the theorem in the case considered. However, the general theorem is an immediate consequence. Since each ideal (other than $\mathfrak{o}$) is (by 4) of the form

$$\mathfrak{a} = \mathfrak{p}_1\mathfrak{p}_2 \cdots \mathfrak{p}_m,$$

where $\mathfrak{p}_1, \mathfrak{p}_2, \ldots, \mathfrak{p}_m$ are prime ideals, it follows that

$$N(\mathfrak{a}) = N(\mathfrak{p}_1)N(\mathfrak{p}_2\mathfrak{p}_3 \cdots \mathfrak{p}_m) = N(\mathfrak{p}_1)N(\mathfrak{p}_2)N(\mathfrak{p}_3 \cdots \mathfrak{p}_m) = \cdots,$$

and hence

$$N(\mathfrak{a}) = N(\mathfrak{p}_1)N(\mathfrak{p}_2) \cdots N(\mathfrak{p}_m).$$

Moreover, if we have

$$\mathfrak{b} = \mathfrak{q}_1\mathfrak{q}_2 \cdots \mathfrak{q}_r,$$

where $\mathfrak{q}_1, \mathfrak{q}_2, \cdots, \mathfrak{q}_r$ again denote prime ideals, then we get

$$\mathfrak{ab} = \mathfrak{p}_1\mathfrak{p}_2 \cdots \mathfrak{p}_m\mathfrak{q}_1\mathfrak{q}_2 \cdots \mathfrak{q}_r,$$

and consequently

$$N(\mathfrak{b}) = N(\mathfrak{q}_1)N(\mathfrak{q}_2)\cdots N(\mathfrak{q}_r),$$
$$N(\mathfrak{ab}) = N(\mathfrak{p}_1)\cdots N(\mathfrak{p}_m)N(\mathfrak{q}_1)\cdots N(\mathfrak{q}_r),$$

which obviously implies

$$N(\mathfrak{ab}) = N(\mathfrak{a})N(\mathfrak{b}).$$

Q.E.D.

8. An ideal $\mathfrak{a}$ (or a number α) is divisible by an ideal $\mathfrak{d}$ (or a number δ) if and only if each power of a prime ideal which divides $\mathfrak{d}$ (or δ) also divides $\mathfrak{a}$ (or α).

Proof. If $\mathfrak{p}$ is a prime ideal and $\mathfrak{p}^m$ is a divisor of an ideal $\mathfrak{d}$ then we have (by 6) $\mathfrak{d} = \mathfrak{d}_1\mathfrak{p}^m$, where $\mathfrak{d}_1$ is an ideal. If we suppose the latter decomposed into all its prime factors, then $\mathfrak{d}$ will also be expressed as a product of prime ideals, among which the factor $\mathfrak{p}$ appears at least m times. Conversely, if the decomposition of $\mathfrak{d}$ into prime factors includes the prime ideal $\mathfrak{p}$ at least m times, then $\mathfrak{d}$ will evidently be divisible by $\mathfrak{p}^m$. Thus if we suppose that every power of the prime ideal which divides $\mathfrak{d}$ also divides an ideal $\mathfrak{a}$, this amounts to saying that all the prime factors in the decomposition of $\mathfrak{d}$ appear, at least as often, as factors in the decomposition of $\mathfrak{a}$. The factors of $\mathfrak{a}$ include first of all the factors of $\mathfrak{d}$ and, if we denote the product of the other factors by $\mathfrak{d}'$, then we have $\mathfrak{a} = \mathfrak{d}\mathfrak{d}'$, and consequently $\mathfrak{a}$ is divisible by $\mathfrak{d}$. The converse proposition, that if $\mathfrak{d}$ is a divisor of $\mathfrak{a}$ then each power of a prime ideal that divides $\mathfrak{d}$ also divides $\mathfrak{a}$, is verified easily. Q.E.D.

If we combine all factors of the same prime in the decomposition of an ideal $\mathfrak{a}$ then we find

$$\mathfrak{a} = \mathfrak{p}^a\mathfrak{q}^b\mathfrak{r}^c\cdots,$$

where $\mathfrak{p}, \mathfrak{q}, \mathfrak{r}, \ldots$ are different prime ideals. By virtue of the preceding theorems we can show that all the divisors $\mathfrak{d}$ of $\mathfrak{a}$ are given by the formula

$$\mathfrak{d} = \mathfrak{p}^{a'}\mathfrak{q}^{b'}\mathfrak{r}^{c'}\cdots,$$

where the exponents $a', b', c', \ldots$ satisfy the conditions

$$0 \le a' \le a, \quad 0 \le b' \le b, \quad 0 \le c' \le c, \quad \ldots.$$

Since any two different combinations of exponents $a', b', c', \ldots$ correspond (by 4) to different ideals $\mathfrak{d}$, the total number of different divisors will be $(a+1)(b+1)(c+1)\cdots$.

9. If $\mathfrak{d}$ is the greatest common divisor of the two ideals $\mathfrak{a}$, $\mathfrak{b}$ then we have

$$\mathfrak{a} = \mathfrak{d}\mathfrak{a}', \quad \mathfrak{b} = \mathfrak{d}\mathfrak{b}',$$

where $\mathfrak{a}'$, $\mathfrak{b}'$ are relatively prime ideals, and the least common multiple $\mathfrak{m}$ of $\mathfrak{a}$, $\mathfrak{b}$ is $\mathfrak{d}\mathfrak{a}'\mathfrak{b}' = \mathfrak{a}\mathfrak{b}' = \mathfrak{b}\mathfrak{a}'$. Moreover, if $\mathfrak{a}\mathfrak{e}$ is divisible by $\mathfrak{b}$, $\mathfrak{e}$ will be divisible by $\mathfrak{b}'$.

We leave the task of finding the proof of this proposition to the reader, along with the rules that allow $\mathfrak{m}$, $\mathfrak{d}$ to be derived from the decompositions of $\mathfrak{a}$, $\mathfrak{b}$ into prime factors.

§26. Congruences

Having established the laws of divisibility for ideals, and hence also for *numbers* in $\mathfrak{o}$, we shall add some considerations on congruences which are important for the theory of ideals. For the moment we shall be content simply to give indications of the proofs.

1. Since $\mathfrak{o}$ is the greatest common divisor of two *relatively prime* ideals $\mathfrak{a}$, $\mathfrak{b}$, and $\mathfrak{a}\mathfrak{b}$ is their least common multiple, then (§2,5) the system of two congruences

$$\omega \equiv \rho \;(\mathrm{mod}\; \mathfrak{a}), \qquad \omega \equiv \sigma \;(\mathrm{mod}\; \mathfrak{b}),$$

where ρ, σ are two given numbers in $\mathfrak{o}$, always has roots ω, and all these roots come under the form

$$\omega \equiv \tau \;(\mathrm{mod}\; \mathfrak{a}\mathfrak{b}),$$

where τ represents a class of numbers modulo $\mathfrak{a}\mathfrak{b}$ which is determined by ρ and σ, or by their corresponding classes modulo $\mathfrak{a}$ and $\mathfrak{b}$ respectively. Conversely, each class τ (mod $\mathfrak{a}\mathfrak{b}$) is determined in this way by precisely one combination ρ (mod $\mathfrak{a}$), σ (mod $\mathfrak{b}$).

We now say that the number ρ is *prime to* the ideal $\mathfrak{a}$ when $\mathfrak{o}\rho$ and $\mathfrak{a}$ are relatively prime ideals, and we let $\psi(\mathfrak{a})$ denote the number of mutually incongruent numbers modulo $\mathfrak{a}$ which are prime to $\mathfrak{a}$. One easily derives the theorem that

$$\psi(\mathfrak{a}\mathfrak{b}) = \psi(\mathfrak{a})\psi(\mathfrak{b})$$

for two relatively prime ideals $\mathfrak{a}$, $\mathfrak{b}$ since τ is prime to $\mathfrak{a}\mathfrak{b}$ if and only if ρ is prime to $\mathfrak{a}$ and σ is prime to $\mathfrak{b}$. Thus we only need to find the value of the function $\psi(\mathfrak{a})$ in the case where $\mathfrak{a}$ is a power $\mathfrak{p}^m$ of the prime ideal $\mathfrak{p}$. The total number of mutually incongruent numbers modulo $\mathfrak{p}^m$ is, in the case $m > 0$, equal to

$$N(\mathfrak{p}^m) = [N(\mathfrak{p})]^m = (\mathfrak{o}, \mathfrak{p}^m) = (\mathfrak{o}, \mathfrak{p})(\mathfrak{p}, \mathfrak{p}^m) = (\mathfrak{p}, \mathfrak{p}^m) N(\mathfrak{p}).$$

It is necessary to subtract from this the number of numbers not prime to $\mathfrak{p}^m$, and hence divisible by $\mathfrak{p}$. Since this number is equal to

$$(\mathfrak{p}, \mathfrak{p}^m) = [N(\mathfrak{p})]^{m-1},$$

we get

$$\psi(\mathfrak{p}^m) = [N(\mathfrak{p})]^m - [N(\mathfrak{p})]^{m-1} = N(\mathfrak{p}^m)\left(1 - \frac{1}{N(\mathfrak{p})}\right),$$

whence we conclude immediately, by virtue of the preceding theorem, that

$$\psi(\mathfrak{a}) = N(\mathfrak{a})\prod\left(1 - \frac{1}{N(\mathfrak{p})}\right)$$

where the product $\prod$ is taken over all the distinct prime ideals $\mathfrak{p}$ that divide the ideal $\mathfrak{a}$. Since we also have

$$\psi(\mathfrak{o}) = 1,$$

we arrive, just as in the theory of rational numbers,[†] at the theorem

$$\sum \psi(\mathfrak{a}') = N(\mathfrak{a}),$$

where the summation is taken over all the ideals $\mathfrak{a}'$ that divide $\mathfrak{a}$.

2. If $\mathfrak{d}$ is the greatest common divisor of the ideals $\mathfrak{a}$ and $\mathfrak{o}\eta$ then we have $\mathfrak{a} = \mathfrak{d}\mathfrak{a}'$, and $\eta\mathfrak{a}'$ will be the least common multiple of $\mathfrak{a}$ and $\mathfrak{o}\eta$, that is, $\mathfrak{a}'$ will be the divisor of $\mathfrak{a}$ corresponding to the number η (§19). Conversely, if $\eta\mathfrak{a}'$ is the least common multiple of $\mathfrak{a}$ and $\mathfrak{o}\eta$ then we have $\mathfrak{a} = \mathfrak{d}\mathfrak{a}'$, where $\mathfrak{d}$ is the greatest common divisor of $\mathfrak{a}$ and $\mathfrak{o}\eta$. It is also clear that the complementary factors $\mathfrak{d}$ and $\mathfrak{a}'$ of the ideal $\mathfrak{a}$ remain the same for all numbers congruent to η modulo $\mathfrak{a}$. It is evident that it will also be the same if we replace η by a number $\eta' \equiv \eta\omega \pmod{\mathfrak{a}}$ where ω is a number prime to $\mathfrak{a}'$. And conversely, if the greatest common divisor $\mathfrak{d}$ of $\mathfrak{a}$, $\mathfrak{o}\eta$ is also that of $\mathfrak{a}$, $\mathfrak{o}\eta'$ then

$$\eta' \equiv \eta\omega, \quad \eta \equiv \eta'\omega' \pmod{\mathfrak{a}},$$

† See Dirichlet's *Vorlesungen über Zahlentheorie*, §14.

whence we deduce

$$\eta\omega\omega' \equiv \eta \ (\text{mod } \mathfrak{a}), \qquad \omega\omega' \equiv 1 \ (\text{mod } \mathfrak{a}'),$$

and consequently ω is a number prime to $\mathfrak{a}'$. Thus the number of mutually incongruent numbers η modulo $\mathfrak{a}$ which correspond to the same divisor $\mathfrak{a}'$ of $\mathfrak{a}$ is $\psi(\mathfrak{a}')$. However, it is necessary to point out that here we have assumed the existence of at least one such number η. Thus if we are given an arbitrary divisor $\mathfrak{a}'$ of the ideal $\mathfrak{a}$ the most we can affirm so far is that the number $\chi(\mathfrak{a}')$ of mutually incongruent numbers η modulo $\mathfrak{a}$ which correspond to the same divisor $\mathfrak{a}'$ will equal either $\psi(\mathfrak{a}')$ or zero. To decide this alternative we consider *all* the incongruent numbers modulo $\mathfrak{a}$, which are $N(\mathfrak{a})$ in number, and partition them into groups of $\chi(\mathfrak{a}')$ numbers, corresponding to the divisors $\mathfrak{a}'$. We must have

$$\sum \chi(\mathfrak{a}') = N(\mathfrak{a})$$

where the summation is taken over the divisors $\mathfrak{a}'$ of $\mathfrak{a}$. But since we also have (1)

$$\sum \psi(\mathfrak{a}') = N(\mathfrak{a}),$$

it follows immediately that $\chi(\mathfrak{a}')$ is never zero and always $\psi(\mathfrak{a}')$. Thus we have proved the following very important theorem:

"If $\mathfrak{d}$ and $\mathfrak{a}'$ are any two ideals we can always, by multiplying $\mathfrak{d}$ by an ideal $\mathfrak{b}'$ prime to $\mathfrak{a}'$, change it into a principal ideal $\mathfrak{d}\mathfrak{b}' = \mathfrak{o}\eta$."

Putting $\mathfrak{d}\mathfrak{a}' = \mathfrak{a}$, the fact that $\psi(\mathfrak{a}')$ is nonzero means that there is always a number η, corresponding to the divisor $\mathfrak{a}'$ of $\mathfrak{a}$, such that $\mathfrak{d}$ will be the greatest common divisor of $\mathfrak{a}$ and $\mathfrak{o}\eta$. If we then put $\mathfrak{o}\eta = \mathfrak{d}\mathfrak{b}'$, $\mathfrak{b}'$ will be an ideal prime to $\mathfrak{a}'$.   Q.E.D.

3. Since each product $\rho\rho'$ of numbers ρ, ρ' prime to an ideal $\mathfrak{a}$ is likewise a number prime to $\mathfrak{a}$, and since, as ρ remains constant and ρ' varies, $\rho\rho'$ runs through a system of $\psi(\mathfrak{a})$ mutually incongruent numbers (mod $\mathfrak{a}$), we deduce by the well-known method,† and for each value of the number ρ, the congruence

$$\rho^{\psi(\mathfrak{a})} \equiv 1 \ (\text{mod } \mathfrak{a}),$$

which represents the highest generalisation of a celebrated theorem of

† See Dirichlet's *Vorlesungen über Zahlentheorie*, §19.

Fermat. For a prime ideal $\mathfrak{p}$ we conclude easily that *every* number ω in the domain $\mathfrak{o}$ satisfies the congruence

$$\omega^{N(\mathfrak{p})} \equiv \omega \ (\mathrm{mod}\ \mathfrak{p}),$$

that is, the congruence

$$\omega^{p^f} \equiv \omega \ (\mathrm{mod}\ \mathfrak{p}),$$

where p is the rational prime number divisible by $\mathfrak{p}$ and f is the degree of the prime ideal $\mathfrak{p}$ (§21,3). This theorem is as important for the theory of the domain $\mathfrak{o}$ as the theorem of Fermat is for the theory of rational numbers, as we shall at least try to make clear by the following remarks, space not permitting us to pursue the general theory further.

If the coefficients of the polynomial function $F(x)$, of degree m, are in $\mathfrak{o}$, and if the coefficient of the highest degree term is not divisible by the prime ideal $\mathfrak{p}$ then we deduce, by well-known reasoning,† that the congruence $F(\omega) \equiv 0 \ (\mathrm{mod}\ \mathfrak{p})$ cannot have more than m mutually incongruent roots. This proposition, combined with the preceding theorem, leads to a complete theory of binomial congruence modulo $\mathfrak{p}$. We deduce, among other things, the existence of *primitive roots* for the prime ideal $\mathfrak{p}$, meaning numbers γ whose powers

$$1, \gamma, \gamma^2, \ldots, \gamma^{N(\mathfrak{p})-2}$$

are mutually incongruent. In general, the theory of congruences of higher degree with rational coefficients carries over completely to functions $F(x)$ whose coefficients are numbers in the domain $\mathfrak{o}$.

However, we have previously ascertained a close connection between the theory of ideals and the theory of higher degree congruences, restricted to the case of *rational* coefficients, developed in the works of Gauss, Galois, Schönemann and Serret.‡ Since all ideals are composed of prime ideals, and each prime ideal $\mathfrak{p}$ divides a determinate rational prime p, we obtain a complete overview of all the ideals in the domain $\mathfrak{o}$ by decomposing all ideals of the form $\mathfrak{o}p$ into their prime factors. The theory of congruences provides a procedure capable of doing this in a great number of cases. If θ is an integer of the field Ω and

$$\Delta(1, \theta, \theta^2, \ldots, \theta^{n-1}) = k^2 \Delta(\Omega)$$

then, *if p is not a divisor of k* we decompose $\mathfrak{o}p$ into prime ideals as

† See Dirichlet's *Vorlesungen über Zahlentheorie*, §26.
‡ See my memoir *Abriss einer Theorie der höheren Congruenzen in Bezug auf einen reellen Primzahl-Modulus* (*Crelle's Journal*, 54).

follows. If $f(t)$ is the polynomial of degree n in t that vanishes for $t = 0$ we can put

$$f(t) \equiv P_1(t)^{a_1} P_2(t)^{a_2} \cdots P_e(t)^{a_e} \quad (\text{mod } p),$$

where $P_1(t), P_2(t), \ldots, P_e(t)$ are distinct irreducible polynomials with respective degrees $f_1, f_2, \ldots, f_e$. Then we certainly have

$$\mathfrak{o}p = \mathfrak{p}_1^{a_1} \mathfrak{p}_2^{a_2} \cdots \mathfrak{p}_e^{a_e},$$

where $\mathfrak{p}_1, \mathfrak{p}_2, \ldots, \mathfrak{p}_e$ are distinct prime ideals with respective degrees $f_1, f_2, \ldots, f_e$. We deduce from this the following extremely important theorem:

"The rational prime p divides the fundamental number $\Delta(\Omega)$ of the field Ω if and only if p is divisible by the square of a prime ideal."

This theorem is still true, although the proof is more difficult, when the numbers k corresponding to all possible numbers θ are all divisible by p. Such a case is actually encountered,† and this is one of the reasons I was determined to build the theory of ideals, not on congruences of higher degree, but on entirely new principles which are at the same time as simple as possible and better suited to the true nature of the subject.

§27. Examples borrowed from circle division

By the general theory of ideals, the fundamentals of which I have developed above, the phenomena of divisibility of numbers in each domain $\mathfrak{o}$, consisting of the integers of a field Ω of finite degree, have been reduced to the same fixed laws that govern the old theory of rational numbers. If we reflect on the infinite variety of these fields Ω, each of which has its own theory of special numbers, it is undoubtedly within the spirit of geometry to ascertain those general laws obeyed by the various theories without exception. However, this is not only of aesthetic or purely theoretical interest, it is also of practical value. The knowledge that these general laws exist greatly assists in the discovery and proof of special phenomena in a given field Ω. To back up this claim to the full we would admittedly need to develop further the general theory of ideals, and to combine it in particular with the algebraic principles of Galois. Instead, I shall simply try to show, using the example that led Kummer

† See the *Göttingische gelehrte Anzeigen* of 20 September 1871, p. 1490.

to introduce his ideal numbers in the first place, how the elements of the theory explained above lead to the goal with great facility.

Let m be a positive rational *prime number*, and let Ω be the field of degree n resulting, in the manner described above (§15), from a primitive root of the equation $\theta^m = 1$, that is, a root of the equation

$$f(\theta) = \theta^{m-1} + \theta^{m-2} + \cdots + \theta^2 + \theta + 1 = 0.$$

Since the coefficients are rational, we have $n \leq m - 1$. Moreover, since $\theta, \theta^2, \ldots, \theta^{m-1}$ are all roots of this equation we have

$$f(t) = \frac{t^m - 1}{t - 1} = (t - \theta)(t - \theta^2) \cdots (t - \theta^{m-1}),$$

where t is a variable, and consequently

$$m = (1 - \theta)(1 - \theta^2) \cdots (1 - \theta^{m-1}).$$

The $m - 1$ factors of the right-hand side are integers and associates of each other, because if r is one of the numbers $1, 2, \ldots, m - 1$ then

$$\frac{1 - \theta^r}{1 - \theta} = 1 + \theta + \theta^2 + \cdots + \theta^{r-1}$$

will be an integer, and if s is positive and chosen so that $rs \equiv 1$ (mod m), then

$$\frac{1 - \theta}{1 - \theta^r} = \frac{1 - \theta^{rs}}{1 - \theta^r} = 1 + \theta^r + \theta^{2r} + \cdots + \theta^{(s-1)r}$$

will also be an integer. Thus if we make the abbreviation

$$1 - \theta = \mu$$

we get

$$m = \epsilon\mu^{m-1}$$

where ϵ is a unit of the field Ω and hence, by taking norms,

$$m^n = [N(\mu)]^{m-1}.$$

However, since m is a prime number, $N(\mu)$ must be a power of m. If we put $N(\mu) = m^e$ it follows that $n = e(m - 1)$, and since $n \leq m - 1$ by the remark above, we conclude that $e = 1$ and $n = m - 1 = \phi(m)$. The preceding equation $f(\theta) = 0$ is therefore *irreducible*, the numbers $\theta, \theta^2, \ldots, \theta^{m-1}$ are conjugate, and these numbers correspond to $m - 1$ isomorphisms transforming the normal field Ω into itself. At the same time we have

$$N(\mu) = m, \quad \mathfrak{o}m = \mathfrak{o}\mu^{m-1}.$$

The principal ideal $\mathfrak{o}\mu$ is a *prime ideal*. In fact, if we have $\mathfrak{o}\mu = \mathfrak{a}\mathfrak{b}$, where $\mathfrak{a}$ and $\mathfrak{b}$ are ideals different from $\mathfrak{o}$, it follows that $m = N(\mathfrak{a})N(\mathfrak{b})$, and since m is a prime number we necessarily have, say, $N(\mathfrak{a}) = m$, $N(\mathfrak{b}) = 1$, whence $\mathfrak{b} = \mathfrak{o}$, contrary to hypothesis. At the same time (§21,3) m is the smallest rational number divisible by μ; the numbers $0, 1, 2, \ldots, m - 1$ form a complete system of mutually incongruent numbers modulo μ. It follows from this that a number of the form

$$\omega = k_0 + k_1\mu + k_2\mu^2 + \cdots + k_{m-2}\mu^{m-2},$$

where $k_0, k_1, k_2, \ldots, k_{m-2}$ are integers, is not divisible by m, and hence also not by μ^{m-1}, unless all the numbers $k_0, k_1, \ldots, k_{m-2}$ are divisible by m. Because an ω divisible by m must also be divisible by μ, and this makes k_0 divisible by μ, and hence also by μ^2, and this makes k_1 divisible by μ, and hence also by m. Continuing in this way, we conclude that the other numbers $k_1, k_2, \ldots, k_{m-2}$ are divisible by m.

With the help of this result it is easy to show that the $m - 1$ numbers $1, \theta, \theta^2, \ldots, \theta^{m-2}$ form a basis for the domain $\mathfrak{o}$ of all integers in the field Ω. Since we have

$$t^m - 1 = (t - 1)f(t), \quad m\theta^{m-1} = (\theta - 1)f'(\theta),$$

it follows, by excluding the uninteresting case $m = 2$, that

$$N[f'(\theta)] = m^{m-2},$$

and since $N(\theta) = 1$ and $N(\theta - 1) = m$ it follows from §17 that

$$\Delta(1, \theta, \theta^2, \ldots, \theta^{m-2}) = (-1)^{\frac{m-1}{2}} m^{m-2}.$$

Moreover, since $\mu = 1 - \theta$, $\theta = 1 - \mu$ it is clear that the two modules $[1, \theta, \ldots, \theta^{m-2}]$ and $[1, \mu, \ldots, \mu^{m-2}]$ are identical, whence it follows (§4,3 and §17,(5)) that we also have

$$\Delta(1, \mu, \mu^2, \ldots, \mu^{m-2}) = (-1)^{\frac{m-1}{2}} m^{m-2}.$$

Since the numbers $1, \mu, \mu^2, \ldots, \mu^{m-2}$ are independent, each number in the field Ω can be put in the form

$$\frac{k_0 + k_1\mu + k_2\mu^2 + \cdots + k_{m-2}\mu^{m-2}}{k} = \frac{\omega}{k}$$

where $k, k_0, k_1, k_2, \ldots, k_{m-2}$ denote rational integers *without common divisor*. For this number to be an integer, that is, for ω to be divisible by k, it is necessary (§18) that k^2 divide the discriminant of the basis $1, \mu, \mu^2, \ldots, \mu^{m-2}$, and therefore k cannot contain prime factors other than the number m. Moreover, since it has been shown that ω is not divisible by m unless the numbers $k_0, k_1, \ldots, k_{m-2}$ are divisible by m, k must also not be divisible by m. Thus we must have $k = \pm 1$, and all integers of the field are of the form

$$\omega = k_0 + k_1\mu + k_2\mu^2 + \cdots + k_{m-2}\mu^{m-2},$$

whence we have

$$\mathfrak{o} = [1, \mu, \ldots, \mu^{m-2}] = [1, \theta, \ldots, \theta^{m-2}],$$

or again, since $1 + \theta + \theta^2 + \cdots + \theta^{m-2} + \theta^{m-1} = 0$,

$$\mathfrak{o} = [\theta, \theta^2, \ldots, \theta^{m-1}], \quad \Delta(\Omega) = (-1)^{\frac{m-1}{2}} m^{m-2}.$$

Now let $\mathfrak{p}$ be any prime ideal different from $\mathfrak{o}\mu$. The positive rational prime p divisible by $\mathfrak{p}$ is different from m, and we have

$$N(\mathfrak{p}) = p^f,$$

where f is the degree of the prime ideal $\mathfrak{p}$. Two powers θ^r, θ^s are congruent modulo such a prime ideal only if they are equal, that is, if $r \equiv s \pmod{m}$. This is because $r \not\equiv s \pmod{m}$ implies $\theta^r - \theta^s = \theta^r(1-\theta^{r-s}) = \epsilon\mu$ where ϵ is a unit, in which case θ^r cannot be congruent to $\theta^s \pmod{\mathfrak{p}}$. Now since we have (§26,3)

$$\theta^{N(\mathfrak{p})} \equiv \theta \pmod{\mathfrak{p}},$$

it follows that

$$p^f \equiv 1 \pmod{m}.$$

Let a be the divisor of $\phi(m) = m - 1$ to which the number p *belongs* modulo m, that is, let a be the least positive exponent for which

$$p^a \equiv 1 \pmod{m}.$$

As we know, f must be divisible by a, and hence $f \geq a$. But since all integers in the field Ω have the form

$$\omega = F(\theta) = x_1\theta + x_2\theta^2 + \cdots + x_{m-1}\theta^{m-1}$$

where $x_1, x_2, \ldots, x_m$ are rational integers, it follows from well-known theorems, true for every prime number p, that

$$\omega^p \equiv F(\theta^p), \quad \omega^{p^r} \equiv F(\theta^{p^r}) \pmod{p},$$

and hence

$$\omega^{p^a} \equiv \omega \pmod{p}.$$

We conclude first of all that the ideal $\mathfrak{o}p$ is a product of *distinct* prime ideals, because if $\mathfrak{o}p = \mathfrak{p}^2\mathfrak{q}$ there would be a number ω divisible by $\mathfrak{p}\mathfrak{q}$ but not by p, so ω^2 and hence also ω^{p^a} would be divisible by $\mathfrak{p}^2\mathfrak{q}^2 = p\mathfrak{q}$, and then also by p, contrary to the preceding congruence. Moreover, since p is divisible by $\mathfrak{p}$, *every* integer ω in the field Ω satisfies the congruence

$$\omega^{p^a} \equiv \omega \pmod{\mathfrak{p}},$$

which therefore has $N(\mathfrak{p}) = p^f$ mutually incongruent roots ω. And since its degree is p^a we must have $p^f \leq p^a$, thus implying $f \leq a$. But it has already been shown that $f \geq a$, so $f = a$. We have therefore arrived at the following result, which is the main theorem of Kummer's theory:[†]

"If p is a prime number different from m and if f is the exponent to which p belongs modulo m, so that $\phi(m) = ef$ for some e, then

$$\mathfrak{o}p = \mathfrak{p}_1\mathfrak{p}_2 \cdots \mathfrak{p}_e,$$

where $\mathfrak{p}_1, \mathfrak{p}_2, \ldots, \mathfrak{p}_e$ are distinct prime ideals of degree f."

The rest follows easily. The general case where m is an arbitrary composite number can be treated similarly. The degree of the normal field Ω is always equal to the number $\phi(m)$ of those numbers among $1, 2, 3, \ldots, m$ that are prime to m. The preceding law is proved without any change, and the determination of the prime ideals that divide m does not present any extra difficulty.

From the very general researches that I am going to publish shortly, the ideals of a normal field Ω immediately allow us to find the ideals of an arbitrary *subfield* [‡] of Ω, that is, any field H whose members all belong to Ω. For example, this enables us to know the ideals of *any* field H resulting from division of the circle and, to give a more precise idea of the scope of these researches, I mention the following case.

Again let m be a prime number, so $\phi(m) = m - 1$, and let e be any divisor of $m - 1 = ef$. In the theory of rational numbers, the congruence

$$k^f \equiv 1 \pmod{m}$$

has precisely f mutually incongruent roots h, which are closed under multiplication and which, in that sense, form a *group*. If θ is again a primitive root of the equation $\theta^m = 1$ and if Ω is the corresponding field of degree $m - 1$, then all the numbers $F(\theta)$ in this field satisfying the

---

[†] Kummer's researches may be found in *Crelle's Journal*, 35, in *Liouville's Journal*, XVI and in the memoirs of the Berlin Academy for the year 1856.

[‡] Dedekind calls it a "divisor", as in his footnote in §15. (Translator's note.)

conditions $F(\theta) = F(\theta^h)$ form a field H of degree e, and the e conjugate periods§ $\eta_1, \eta_2, \ldots, \eta_e$ are sums of f terms, one of which is

$$\eta = \sum \theta^h,$$

and they form a basis of the domain $\mathfrak{e}$ of integers in H. By general results I shall speak of later (or else immediately, by conclusions like those derived above in the case $e = m - 1$) we now obtain the prime ideals belonging to this subfield H of the normal field Ω. If we put

$$\rho = \prod (1 - \theta^h),$$

then ρ is an integer of the field H, m is an associate of ρ^e, and $\mathfrak{e}\rho$ is a prime ideal. Moreover, if p is a rational prime different from m, and if p^f belongs to the exponent f' modulo m, then f' will necessarily be a divisor of $e = e'f'$ and the principal ideal $\mathfrak{e}p$ will be the product of e' distinct prime ideals of degree f'. In the case $e = m - 1$, $f = 1$, H is identical with Ω, and we again obtain the result proved above. We now examine more closely the case $e = 2$, $f = \frac{m-1}{2}$.

In this case the f numbers h are the quadratic residues of m. If we let k denote a quadratic nonresidue, then the two conjugate periods

$$\eta = \sum \theta^h, \quad \eta' = \sum \theta^k$$

form a basis of the domain $\mathfrak{e}$ of all integers in the quadratic field H, and hence its discriminant will be

$$\Delta(H) = \begin{vmatrix} \eta & \eta' \\ \eta' & \eta \end{vmatrix}^2 = (\eta - \eta')^2,$$

because $\eta + \eta' = -1$. The number m is an associate of the square of the number $\rho = \prod(1 - \theta^k)$, and $\mathfrak{e}\rho$ is a prime ideal. Moreover, $\mathfrak{e}p$ is the product of two prime ideals of degree 1, or else $\mathfrak{e}p$ is a prime ideal of degree 2, according as

$$p^{\frac{m-1}{2}} \equiv +1 \text{ or } -1 \pmod{m},$$

that is, using the notation of Legendre, according as

$$\left(\frac{p}{m}\right) = +1 \text{ or } -1.$$

But we can directly study all quadratic fields, without recourse to division of the circle, and we have already (§18) determined the discriminant D' of such a field H. From D' we can also derive the prime ideals† of

§ *Disquisitiones Arithmeticae*, art. 343.
† See Dirichlet's *Vorlesungen über Zahlentheorie*, §168.

the field H. If the rational prime p divides D' then the corresponding principal ideal $\mathfrak{e}p$ will be the square of a prime ideal. If p does not divide D', and if p is odd, then $\mathfrak{e}p$ will be the product of two prime ideals of degree 1, or else a prime ideal of degree 2, according as

$$\left(\frac{D'}{p}\right) = +1 \text{ or } -1.$$

Finally, if D' is odd and hence $\equiv 1 \pmod 4$ then $\mathfrak{e}(2)$ will be the product of two prime ideals of degree 1, or else a prime ideal of degree 2, according as

$$D' \equiv 1 \text{ or } 5 \pmod 8.$$

Comparing these laws, valid for all quadratic fields, with the result for the special field H derived using circle division, we see first that D' must be divisible by m, but not by any other prime number, and hence that (§18)

$$\Delta(H) = D' = (-1)^{\frac{m-1}{2}} m.$$

In this way we derive, from entirely general principles and without calculation, the result

$$(\eta - \eta')^2 = (-1)^{\frac{m-1}{2}} m,$$

previously demonstrated in the theory of circle division by effective formation of the square of $\eta - \eta'$.[†] Pursuing this comparison further, we are led again to the theorem

$$\left(\frac{p}{m}\right) = \left(\frac{\pm m}{p}\right),$$

where $\pm m \equiv 1 \pmod 4$, and to the theorem

$$\left(\frac{2}{m}\right) = (-1)^{\frac{m^2-1}{8}}.$$

This proof of the quadratic reciprocity law, in which we also determine the quadratic character of the number -1, is essentially the same as the celebrated sixth proof of Gauss, [‡] later reproduced in different forms by Jacobi, Eisenstein and others. I should say that it was by meditating on the essence of that proof and the analogous proofs of cubic and biquadratic reciprocity that I was led to the general researches mentioned above and soon to be published.

† *Disquisitiones Arithmeticae*, art. 356.
‡ *Theorematis fundamentalis in doctrina de residuis quadraticis demonstrationes et ampliationes novæ*, 1817.

As a final example, we consider the case $m = 4$. We then have $\theta = i = \sqrt{-1}$ and the integers of the quadratic field Ω are the complex integers, first introduced by Gauss, of the form

$$\omega = x + yi,$$

where x, y are rational integers (§6). The discriminant of the field is

$$\begin{vmatrix} 1 & i \\ 1 & -i \end{vmatrix}^2 = -4$$

The number $2 = i(i - 1)^2$ is an associate of the square of the prime number $1 - i$. If p is a positive odd rational prime then we have

$$i^p = (-1)^{\frac{p-1}{2}} i,$$

and consequently

$$\omega^p = (x + yi)^p \equiv x + (-1)^{\frac{p-1}{2}} yi \pmod{p}.$$

Now if $p \equiv 1 \pmod 4$, each integer ω will satisfy the congruence

$$\omega^p \equiv \omega \pmod{p},$$

whence it follows immediately that $\mathfrak{o}p$ is the product of two different prime ideals of degree 1. But if $p \equiv 3 \pmod 4$ we have

$$\omega^p \equiv \omega', \quad \omega^{p^2} \equiv \omega \pmod{p},$$

where ω' is the number conjugate to ω, and we conclude easily that $\mathfrak{o}p$ is a prime ideal of degree 2. But every ideal $\mathfrak{a}$ of this field must be a principal ideal. In fact, if α_0 is a member of the ideal $\mathfrak{a}$ with *minimum* norm, then each number α in the ideal will be divisible by α_0. This is because (§6) we can choose the integer ω so that

$$N(\alpha - \omega\alpha_0) < N(\alpha_0),$$

and since the numbers α, α_0, and hence also $\alpha - \omega\alpha_0$, belong to the ideal $\mathfrak{a}$, we must have $N(\alpha - \omega\alpha_0) = 0$, whence $\alpha = \omega\alpha_0$ and consequently $\mathfrak{a} = \mathfrak{o}\alpha_0$.   Q.E.D.

Now, since $\mathfrak{o}p$ is the product of two prime ideals of degree 1 in the case where p is a rational prime $\equiv 1 \pmod 4$, it follows that

$$p = N(\alpha_0) = N(a + bi) = a^2 + b^2,$$

which is the celebrated theorem of Fermat.

§28. Classes of ideals

We now return to consideration of an arbitrary field Ω of degree n, in order to establish the distribution of its ideals into *classes*. This distribution depends first of all on the theorem (§25,5) that each ideal $\mathfrak{a}$ can be converted into a principal ideal by multiplication by an ideal $\mathfrak{m}$, and on the following definition: two ideals $\mathfrak{a}$, $\mathfrak{a}'$ are called *equivalent* when they can be converted into principal ideals $\mathfrak{a}\mathfrak{m} = \mathfrak{o}\mu$, $\mathfrak{a}'\mathfrak{m} = \mathfrak{o}\mu'$ by multiplication by the same ideal $\mathfrak{m}$. Then we evidently have $\mu'\mathfrak{a} = \mu\mathfrak{a}'$ and, conversely, if there are two nonzero numbers η, η' such that $\eta'\mathfrak{a} = \eta\mathfrak{a}'$ then the ideals $\mathfrak{a}$, $\mathfrak{a}'$ are certainly equivalent. Because, multiplying $\mathfrak{a}$ by $\mathfrak{m}$ to obtain a principal ideal $\mathfrak{a}\mathfrak{m} = \mathfrak{o}\mu$, it follows that $\mathfrak{o}\mu\eta' = \eta'\mathfrak{a}\mathfrak{m} = \eta\mathfrak{a}'\mathfrak{m}$. Then $\mu\eta'$ is divisible by η, whence $\mu\eta' = \mu'\eta$, $\mathfrak{o}\mu'\eta = \eta\mathfrak{a}'\mathfrak{m}$, hence $\mathfrak{a}'\mathfrak{m} = \mathfrak{o}\mu'$.   Q.E.D.

If two ideals $\mathfrak{a}'$, $\mathfrak{a}''$ are equivalent to a third, $\mathfrak{a}$, then $\mathfrak{a}'$ and $\mathfrak{a}''$ are equivalent to each other. This is because the hypothesis yields four numbers μ, μ', η, η'' such that $\mu'\mathfrak{a} = \mu\mathfrak{a}'$, $\eta''\mathfrak{a} = \eta\mathfrak{a}''$, and hence $(\eta''\mu)\mathfrak{a}' = (\mu'\eta)\mathfrak{a}''$. Q.E.D.

It follows that the ideals can be partitioned into classes. If $\mathfrak{a}$ is a given ideal, the system A of all ideals $\mathfrak{a}, \mathfrak{a}', \mathfrak{a}'', \ldots$ equivalent to $\mathfrak{a}$ will be called a *class of ideals*, and $\mathfrak{a}$ will be called a *representative* of the class A. Any two ideals in A are equivalent, and any ideal $\mathfrak{a}'$ in A can be chosen as a representative in place of $\mathfrak{a}$.

It is clear that the system of all principal ideals forms a class by itself, since each of them is converted to itself when multiplied by the ideal $\mathfrak{o}$, and hence they are all equivalent. And if an ideal $\mathfrak{a}$ is equivalent to a principal ideal, and hence equivalent to $\mathfrak{o}$, then $\mathfrak{a}$ itself must be a principal ideal, since there are two numbers μ, μ' such that $\mu'\mathfrak{a} = \mathfrak{o}\mu$ and this implies that μ is divisible by μ', whence $\mu = \mu'\mu''$ and consequently $\mathfrak{a} = \mathfrak{o}\mu''$. Thus the class represented by $\mathfrak{o}$ includes all principal ideals and no others. We call this class the *principal class* and denote it by O.

Now if $\mathfrak{a}$ runs through all the ideals in a class A, and $\mathfrak{b}$ runs through all the ideals in a class B, then all the products $\mathfrak{a}\mathfrak{b}$ belong to the same class K. Because if $\mathfrak{a}'$, $\mathfrak{a}''$ are in A and $\mathfrak{b}'$, $\mathfrak{b}''$ are in B there are four numbers α', α'', β', β'' such that $\alpha''\mathfrak{a}' = \alpha'\mathfrak{a}''$, $\beta''\mathfrak{b}' = \beta'\mathfrak{b}''$, and it follows that $(\alpha''\beta'')(\mathfrak{a}'\mathfrak{b}') = (\alpha'\beta')(\mathfrak{a}''\mathfrak{b}'')$, that is, $\mathfrak{a}'\mathfrak{b}'$ and $\mathfrak{a}''\mathfrak{b}''$ are equivalent ideals. We denote the class K to which all the products $\mathfrak{a}\mathfrak{b}$ belong by AB and call it the *product* of A and B, or the class *composed* from A and B. We evidently have $AB = BA$, and it follows from the equation

$(\mathfrak{ab})\mathfrak{c} = \mathfrak{a}(\mathfrak{bc})$ that $(AB)C = A(BC)$ for any three classes A, B, C. Then we can apply the same reasoning as for the multiplication of numbers or ideals, and show that, in the composition of any number of classes $A_1, A_2, \ldots, A_m$, the order in which pairs of classes are combined has no influence on the final result, which we can denote simply by $A_1 A_2 \cdots A_n$. If the ideals $\mathfrak{a}_1, \mathfrak{a}_2, \ldots, \mathfrak{a}_m$ represent the classes $A_1, A_2, \ldots, A_m$ then the ideal $\mathfrak{a}_1 \mathfrak{a}_2 \cdots \mathfrak{a}_m$ represents the class $A_1 A_2 \cdots A_m$. If all the m factors equal A, then their product is called the mth power of A, and we denote it by A^m. In addition, we put $A^1 = A$ and $A^0 = O$. The following two cases are particularly important:

The equation $\mathfrak{oa} = \mathfrak{a}$ yields the theorem that $OA = A$ for any class A.

Moreover, since each ideal $\mathfrak{a}$ can be transformed into a principal ideal $\mathfrak{am}$ by multiplication by an ideal $\mathfrak{m}$, for each class A there is a class M satisfying the condition $AM = O$, and only one, because if the class N is such that $AN = O$ it follows that

$$N = NO = N(AM) = M(AN) = MO = M.$$

The class M is called the class *opposite* or *inverse* to A, and we denote it by A^{-1}. Conversely, it is clear that A will be the class inverse to A^{-1}. If in addition we define A^{-m} to be the class inverse to A^m then we have the following theorems for any rational integer exponents r, s:

$$A^r A^s = A^{r+s}, \quad (A^r)^s = A^{rs}, \quad (AB)^r = A^r B^r.$$

Finally, it is evident that from $AB = AC$ we can always deduce $B = C$, multiplying by A^{-1}.

§29. *The number of classes of ideals*

If we take any n integers $\omega_1, \omega_2, \ldots, \omega_n$ forming a basis for the field Ω, then each number

$$\omega = h_1 \omega_1 + h_2 \omega_2 + \cdots + h_n \omega_n$$

with rational integer coordinates $h_1, h_2, \ldots, h_n$ will be an integer in the same field. If we allow the coordinates to take all integer values of absolute value not greater than a particular positive value k, then it is evident that the absolute values of the corresponding numbers ω, which are real, or their analytic moduli, which are imaginary, are all $\leq rk$ where r is the sum of the absolute values or moduli of $\omega_1, \omega_2, \ldots, \omega_n$, and hence a constant independent of k. Moreover, since the norm $N(\omega)$ is a product of n conjugate numbers ω of the form above, we also have

$$\pm N(\omega) \le sk^n,$$

where s is likewise a constant depending only on the basis. We deduce from this the following theorem:

In each class M of ideals there is at least one ideal $\mathfrak{m}$ whose norm is bounded by a constant.

Proof. Take any ideal $\mathfrak{a}$ in the inverse class M^{-1}, and let k be the positive rational integer determined by the conditions

$$k^n \le N(\mathfrak{a}) < (k+1)^n.$$

If we now allow each of the n coordinates $h_1, h_2, \ldots, h_n$ to take all $k+1$ values $0, 1, 2, \ldots, k$, then we obtain distinct numbers ω and, since their number is $(k+1)^n$ and hence $> N(\mathfrak{a})$, there are necessarily two different numbers ω,

$$\beta = b_1 \omega_1 + \cdots + b_n \omega_n, \quad \gamma = c_1 \omega_1 + \cdots + c_n \omega_n,$$

which are congruent modulo $\mathfrak{a}$. Hence their difference

$$\alpha = (b_1 - c_1)\omega_1 + \cdots + (b_n - c_n)\omega_n$$

will be a nonzero number divisible by $\mathfrak{a}$. But since the coordinates b, c of the numbers β, γ come from the sequence $0, 1, 2, \ldots, k$, the coordinates $b - c$ of α all have absolute value not greater than k, and hence

$$\pm N(\alpha) \le sk^n.$$

Since α is divisible by $\mathfrak{a}$ we have $\mathfrak{o}\alpha = \mathfrak{a}\mathfrak{m}$, where $\mathfrak{m}$ is an ideal in the class M, and hence

$$\pm N(\alpha) = N(\mathfrak{a})N(\mathfrak{m}) \le sk^n.$$

Moreover, since $k^n \le N(\mathfrak{a})$, it follows that $N(\mathfrak{m}) \le s$.   Q.E.D.

If we now consider that the norm m of an ideal $\mathfrak{m}$ is always divisible by $\mathfrak{m}$ (§20), it is clear that there cannot be more than a finite number of ideals $\mathfrak{m}$ with given norm m, because each ideal, and in particular $\mathfrak{o}m$, is divisible by only a finite number of ideals (§25,8). Since there are also only a finite number of rational integers m not exceeding a given constant s, there cannot be more than a finite number of ideals $\mathfrak{m}$ with $N(\mathfrak{m}) \le s$, which evidently yields the fundamental theorem:

The number of classes of ideals of the field Ω is finite.

The *exact* determination of the number of classes of ideals is incontestably a very important problem, but also one of the most difficult in the theory of numbers. For quadratic fields, whose theory essentially coincides with that of binary quadratic *forms*, we know that the problem was first solved by Dirichlet.† His solution, expressed in the terminology of the theory of *ideals*, rests on the study of the function

$$\sum \frac{1}{N(\mathfrak{a})^s} = \prod \frac{1}{1 - \frac{1}{N(\mathfrak{p})^s}},$$

for infinitely small positive values of the independent variable $s - 1$. The sum is taken over all ideals $\mathfrak{a}$, the product over all prime ideals $\mathfrak{p}$, and the identity of the two expressions is an immediate consequence of the laws of divisibility. With the aid of these principles, the number of classes of forms or ideals has been later determined by Eisenstein† for a particular case of a field of degree 3, and by Kummer‡ for the higher degree fields arising from division of the circle. These researches have excited the liveliest interest because of their astonishing connections with analysis, algebra and other parts of number theory. For example, the problem treated by Kummer is closely related to Dirichlet's proof of the theorem on primes in arithmetic progressions, which can be considerably simplified with the aid of these researches. There is no doubt that further study of the general problem will lead to important progress in these branches of mathematics; however, while part of this research has been successfully completed for an arbitrary field Ω,§ we are nevertheless far from the complete solution, and for the moment we are confined to studying new special cases.

§30. Conclusion

We shall derive some further interesting consequences of the fundamental theorem proved above. (See *Disquisitiones Arithmeticae*, art. 305-307.)

Let h be the number of classes of ideals of the field Ω, and let A be a particular class. The $h + 1$ powers

$$O, A, A^2, \ldots, A^{h-1}, A^h$$

† *Crelle's Journal*, 19, 21.
† *Crelle's Journal*, 28.
‡ *Crelle's Journal*, 40, *Liouville's Journal*, XVI.
§ Dirichlet, *Vorlesungen über Zahlentheorie*, §167.

cannot all be different, hence there are two different exponents r and $r + m > r$ in the sequence $0, 1, 2, \ldots, h$ such that $A^{r+m} = A^r$, and consequently

$$A^m = O.$$

Moreover, if m is the *smallest* positive exponent satisfying this condition it is easy to see that the m classes

$$O, A, A^2, \ldots, A^{m-1}$$

are all different, and we say that the class A *belongs* to the exponent m. Obviously $A^{m-1} = A^{-1}$, and more generally we have $A^r = A^s$ if and only if $r \equiv s \pmod{m}$. Then, if B denotes any class, the m classes

$$(B) \qquad\qquad B, BA, BA^2, \ldots, BA^{m-1}$$

will be all different, and any two complexes of m classes, such as the preceding (B) and the following

$$(C) \qquad\qquad C, CA, CA^2, \ldots, CA^{m-1},$$

will either be identical or have nothing in common. In fact, if both include the same class $BA^r = CA^s$ then we have $C = BA^{r-s}$, whence it follows immediately that the m classes in (C) are the same as those in (B). Thus the whole system of h classes is partitioned into g such complexes and, since each complex includes m different classes, we have $h = mg$. That is, the exponent m to which the class A belongs is always a divisor of the number of classes, h. It follows that we have the theorem

$$A^h = O$$

for each class A. Now if $\mathfrak{a}$ is any ideal in any class A, then $\mathfrak{a}^h$ belongs to the class A^h, and hence to the principal class. That is, the hth power of any ideal is a principal ideal.

With this important theorem we come to see the notion of *ideal* from a new point of view, at the same time connected with a precise definition of *ideal numbers*. Let $\mathfrak{a}$ be any ideal and let $\mathfrak{a}^h = \mathfrak{o}\alpha_1$. Now if α denotes any number in the ideal $\mathfrak{a}$, α^h will be in $\mathfrak{a}^h$, and hence divisible by the number α_1, and it follows from §13,2 that α is divisible by the integer $\mu = \sqrt[h]{\alpha_1}$, which does not in general belong to the field Ω. Conversely, if α is an integer belonging to Ω and divisible by μ, then α^h will be divisible by $\mu^h = \alpha_1$ and consequently $(\mathfrak{o}\alpha)^h$ will be divisible by $\mathfrak{o}\alpha_1 = \mathfrak{a}^h$, from which we easily conclude, by the general laws of divisibility (§25), that $\mathfrak{o}\alpha$ is divisible by α, so that α is a number in the ideal $\mathfrak{a}$. Thus the ideal a *consists of* all the integers in Ω divisible by the integer μ. For this reason we say that the number μ, though not actually a member of Ω, is an *ideal number of the field* Ω, and that it *corresponds* to the ideal a. Or, a little more generally, an algebraic integer μ is said to be an ideal number of the field Ω when there is a power μ^r, with positive integer exponent r, equal to an *actual* η in Ω, and at the same time an ideal a in the field Ω satisfies the condition $a^r = o\mu$. The latter ideal a is the ideal corresponding to the ideal number μ, and it is a principal ideal if and only if μ is an associate of an actual number in the field Ω. (See the Introduction and §10.)

We end our considerations with the proof of the following theorem announced earlier (§14):

Any two algebraic integers α, β have a common divisor δ which can be expressed in the form $\delta = \alpha\alpha' + \beta\beta'$, where α' and β' are likewise algebraic integers.

Proof. We assume that the two numbers α, β are nonzero, otherwise the theorem is evident. Then it is easy to see that there is a field Ω of finite degree including both α, β. Let o be the domain of integers of this field, and let h be the number of classes of ideals. Now put

$$o\alpha = a\mathfrak{d}, \quad o\beta = b\mathfrak{d}, \quad \mathfrak{d}^h = o\delta_1,$$

where $\mathfrak{d}$ is the greatest common divisor of $o\alpha$, $o\beta$, and δ_1 is in o. Since α^h, β^h are divisible by $\mathfrak{d}^h$, we can put

$$\alpha^h = \alpha_1\delta_1, \quad \beta^h = \beta_1\delta_1, \quad o\alpha_1 = a^h, \quad o\beta_1 = b^h,$$

where α_1, β_1 are likewise in o. Also, since a and b are relatively prime ideals, o will be the greatest common divisor of $o\alpha_1$, $o\beta_1$ and, since the number 1 is in o, there will be two numbers α_2, β_2 in o satisfying the condition

$$\alpha_1\alpha_2 + \beta_1\beta_2 = 1, \quad \text{or} \quad \alpha^h\alpha_2 + \beta^h\beta_2 = \delta_1.$$

If we now put

$$\delta_1 = \delta^h,$$

then the integer δ will be a common divisor of α and β, since α^h, β^h are divisible by δ_1, and hence, since $h \geq 1$, we can put

$$\alpha_2\alpha^{h-1} = \alpha'\delta^{h-1}, \quad \beta_2\beta^{h-1} = \beta'\delta^{h-1},$$

where α', β' are integers satisfying the condition $\alpha\alpha' + \beta\beta' = \delta$. Q.E.D.

If at least one of the two numbers α, β is nonzero, then the number δ, and any of its associates, deserves the name *greatest* common divisor of α, β. If δ is a unit then α, β may be called *relatively prime*, and two such numbers enjoy the characteristic property that any number μ divisible by α and β is also divisible by $\alpha\beta$. This is because the equations $\mu = \alpha\alpha'' = \beta\beta''$ and $1 = \alpha\alpha' + \beta\beta'$ imply

$$\mu = \alpha\beta(\alpha'\beta'' + \beta'\alpha''),$$

and the converse is equally valid, since α, β are both nonzero.

Index

153

154   *Index*

root
  of congruence, 137
  of unity, 28, 38
  primitive, 33, 137, 139, 142

Schönemann, 137
section, 44, 58
Serret, 137
Stark, 42
subfield, 142
substitution, 108
  inverse, 13
  unimodular, 14
symmetric functions, 41, 113
  Newton theorem, 40

two square theorem, 9
  and Gaussian primes, 24
  Dedekind proof, 25, 145
  Euler proof, 11
  Fermat proof, 11
  Lagrange proof , 15

unimodular substitution, 14
unique prime factorisation
  and equivalence of forms, 28
  failure in $\mathbb{Z}[\sqrt{-3}]$, 30
  failure in $\mathbb{Z}[\sqrt{-5}]$, 5, 27
  failure in $\mathbb{Z}[\zeta_{23}]$, 32
  failure in cyclotomic integers, 56
  failure in quadratic field, 87
  for ideals, 3, 130
  in complex integers, 56
  in *Disquisitiones*, 7
  in Gaussian integers, 22, 24, 85
  in rational integers, 22, 56, 84
  in $\mathbb{Z}[\sqrt{-2}]$, 26
  in $\mathbb{Z}[\zeta_3]$, 30
  of ideals, 5, 102
units
  in algebraic integers, 54, 106
  in Gaussian integers, 23, 85
  in quadratic field, 86
  in rational integers, 83

vector space, 41

Weber, 45
  and Dedekind, 46
Weil, 13

zero module, 63